**Organic Principles and Practices Handbook Series**
A Project of the Northeast Organic Farming Association

# Compost, Vermicompost, and Compost Tea

*Feeding the Soil on the Organic Farm*

**Revised and Updated**

GRACE GERSHUNY

Illustrated by Jocelyn Langer

CHELSEA GREEN PUBLISHING
WHITE RIVER JUNCTION, VERMONT

Editorial Coordinator: Makenna Goodman
Project Manager: Bill Bokermann
Copy Editor: Cannon Labrie
Proofreader: Helen Walden
Indexer: Peggy Holloway
Designer: Peter Holm, Sterling Hill Productions

Printed in the United States of America
First Chelsea Green revised printing March, 2011
10 9 8 7 6 5 4 3 2 1   11 12 13 14

**Our Commitment to Green Publishing**
Chelsea Green sees publishing as a tool for cultural change and ecological stewardship. We strive to
align our book manufacturing practices with our editorial mission and to reduce the impact of our
business enterprise in the environment. We print our books and catalogs on chlorine-free recycled
paper, using vegetable-based inks whenever possible. This book may cost slightly more because we use
recycled paper, and we hope you'll agree that it's worth it. Chelsea Green is a member of the Green
Press Initiative (www.greenpressinitiative.org), a nonprofit coalition of publishers, manufacturers, and
authors working to protect the world's endangered forests and conserve natural resources. *Compost,
Vermicompost, and Compost Tea* was printed on Joy White, a 30-percent postconsumer recycled paper
supplied by Thomson-Shore.

Library of Congress Cataloging-in-Publication Data
Gershuny, Grace.
 Compost, vermicompost, and compost tea : feeding the soil on the organic farm / Grace Gershuny ;
illustrated by Jocelyn Langer. -- Updated.
    p. cm. -- (Organic principles and practices handbook series)
"A Project of the Northeast Organic Farming Association."
Includes bibliographical references and index.
ISBN 978-1-60358-347-3
1. Compost--United States. 2. Vermicomposting--United States. 3. Compost tea--United States. I.
Langer, Jocelyn. II. Northeast Organic Farming Association. III. Title. IV. Title: Feeding the soil on the
organic farm. V. Series: Organic principles and practices handbook series.
  S661.G45 2011
  631.8'75--dc22
                                    2010053923

Chelsea Green Publishing Company
Post Office Box 428
White River Junction, VT 05001
(802) 295-6300
www.chelseagreen.com

## Best Practices for Farmers and Gardeners

The NOFA handbook series is designed to give a comprehensive view of key farming practices from the organic perspective. The content is geared to serious farmers, gardeners, and homesteaders and those looking to make the transition to organic practices.

Many readers may have arrived at their own best methods to suit their situations of place and pocketbook. These handbooks may help practitioners review and reconsider their concepts and practices in light of holistic biological realities, classic works, and recent research.

Organic agriculture has deep roots and a complex paradigm that stands in bold contrast to the industrialized conventional agriculture that is dominant today. It's critical that organic farming get a fair hearing in the public arena—and that farmers have access not only to the real dirt on organic methods and practices but also to the concepts behind them.

## About This Series

The Northeast Organic Farming Association (NOFA) is one of the oldest organic agriculture organizations in the country, dedicated to organic food production and a safer, healthier environment. NOFA has independent chapters in Connecticut, Massachusetts, New Hampshire, New Jersey, New York, Rhode Island, and Vermont.

This handbook series began with a gift to NOFA/Mass and continues under the NOFA Interstate Council with support from NOFA/Mass and a generous grant from Sustainable Agriculture Research and Education (SARE). The project has utilized the expertise of NOFA members and other organic farmers and educators in the Northeast as writers and reviewers. Help also came from the Pennsylvania Association for Sustainable Agriculture and from the Maine Organic Farmers and Gardeners Association.

Jocelyn Langer illustrated the series, and Jonathan von Ranson edited it and coordinated the project. The Manuals Project Committee included Bill Duesing, Steve Gilman, Elizabeth Henderson, Julie Rawson, and Jonathan von Ranson. The committee thanks SARE and the wonderful farmers and educators whose willing commitment it represents.

# CONTENTS

# Compost Poetry

## On Top

All this new stuff goes on top
Turn it over, turn it over
Wait and water down
From the dark bottom
Turn it inside out
Let it spread through
Sift down even.
Watch it sprout.

A mind like compost.
—GARY SNYDER

## Kneeling Here, I Feel Good

Sand: crystalline children
Of dead mountains.
Little quartz worlds
Rubbed by the wind.

Compost: rich as memory,
Sediment of our pleasures,
Orange rinds and roses and beef bones,
Coffee and cork and dead lettuce,
Trimmings of hair and lawn.

I marry you, I marry you.
In your mingling under my grubby nails
I touch the seeds of what will be.

Revolution and germination
Are mysteries of birth
Without which
Many
Are born to starve.

I am kneeling and planting.
I am making fertile.
I am putting
Some of myself
Back in the soil.
Soon enough
Sweet black mother of our food
You will have the rest.

—MARGE PIERCY

# one

# Why Compost?

Composting represents, for many, the essence of organic food production. Aside from avoiding toxic agrichemicals, composting is what sets organic methods apart from conventional ones. High-quality compost consists primarily of humus—the fragrant, spongy, nutrient-rich material resulting from decomposition of organic matter—and offers the same benefits as nonorganic conventional methods: it creates and supports the biological processes in the soil. Compost is a microbe-laden substance that inoculates the soil with diverse beneficial organisms. It is a source of organic matter as well as carrying a modest mineral fertilizer value.

Many people don't think that the extra time and expense needed to make compost is necessary, since the humification process takes place naturally when raw organic wastes are incorporated into the soil (known as sheet composting). However, only active composting can guarantee humus as the end product, imparting a quality rarely attained in sheet composting. Moreover, many raw materials won't readily decompose in the soil. It's a rare soil that is healthy enough to buffer the nitrogen hit or pH impact of raw manure, break down carbonaceous materials such as sawdust, and digest such wet materials as cannery wastes.

One study compared compost with stockpiled feedlot manure in the Midwest. Compost-treated plots produced yields similar to those amended with four times the amount of manure. Soil-quality indicators (pH, organic matter, cation exchange capacity, and major nutrient levels) as well as plant tissue (with regard to nitrogen, phosphorus, and potassium) showed greater improvement in the compost plots. These differences cannot be accounted for by the actual nutrient content of the compost, indicating that it's probably their increased availability that makes the difference.

Composting does involve extra time and money, but offers the following practical advantages in addition to the benefits described above:

- It stabilizes the volatile nitrogen fraction by fixing it into organic forms (usually the bodies of microbes).
- It allows use of materials that may be toxic to soil organisms, such as cannery wastes, or that will steal nitrogen if applied raw, such as sawdust.
- It permits an even distribution of trace minerals, avoiding the problem of spot imbalances.
- It eliminates most objectionable odors created by bacterial action on sulfur and nitrogen compounds.
- It reduces the volume of wastes, and therefore the number of trips over the field.
- It eliminates most pathogens and weed seeds thermally, and reduces the presence of human pathogens such as *E. coli*.
- The final product is easy to store and handle, and versatile in its applications.
- Compost and compost tea have been shown to offer significant anti-fungal disease benefits. (Gershuny and Smillie, 92–93)

In spite of these benefits, many organic farmers do not compost, and consider it to be more bother than it's worth. Some may not see the need to compost. Farmers who don't keep livestock or who don't have access to significant amounts of livestock manure may not think they have enough raw organic materials available to do so. Moreover, alternatives beckon: commercial fertilizer materials that are approved for organic production, rotation plans that include copious quantities of green manures, and direct application of manure and/or other types of raw organic materials—all are alternatives to making compost on the farm. Finally, many munici-palities and private enterprises produce compost (of varying quality levels) in industrial quantities (at varying prices), and make it available to farmers either bagged or in bulk.

Given these choices, what are the advantages of making one's own compost? Is there really any benefit to using compost as opposed to build-ing soil organic matter through other means? How can the decision about compost making affect a farm's economics? This manual will help answer these questions, and is specifically intended to address the conditions faced by organic farmers in the Northeast. It also offers information about the basic nuts and bolts of farm-scale composting, describes some of the methods, including vermicompost (using worms) and compost tea, used by successful compost makers in our region, and lists resources for developing a compost system that works for you.

A few professional northeastern composters—Arnie Voehringer, Mike Merner, Karl Hammer, and Mike Lombard—were kind enough to share information about their methods and materials, as well as the economics of their operations. Their information is included throughout the text to illustrate the range of creative approaches that may profitably be used to compost on a commercial scale in our region. Arnie Voehringer has been composting since 1976, when he was the farm manager at the Rodale Research Center. He is now in charge of composting at White Oak Farm, affiliated with the New England Small Farms Institute in Belchertown, Massachusetts. Mike Merner has been composting professionally for over thirty years at Earth Care Farm in Charlestown, Rhode Island. Karl Hammer was formerly the owner of Moody Hill Farms in New York, and then returned to Montpelier, Vermont, to resume commercial-scale composting as the Vermont Compost Company. Mike Lombard has operated Ideal Compost in Peterborough, New Hampshire, for about twenty years.

# two

# Principles and Biology of Composting

Composting is the art and science of mixing various organic materials in a pile, monitoring the resultant biological activity, and controlling conditions so that the original raw substances are transformed into stable humus. This process of humification is a combination of biochemical degradation and microbial synthesis. Composting is a form of aerobic digestion or controlled fermentation, and differs markedly from anaerobic breakdown, or putrefaction.

Complex chemical and biological processes are involved in transforming raw organic material into finished compost. Compost ecosystems consist of vast numbers and types of organisms, whose interactions are only slightly known. Different organisms predominate at different stages of decomposition, in different climates and soil types, at different pH, oxygen, and temperature levels, on different kinds of organic feedstocks, and at different times of the year. Macrofauna, such as mites, millipedes, springtails, beetles, ants, flies, nematodes, and most importantly, earthworms, set the stage by physically breaking down the raw materials through biting, chewing, sucking, and grinding. Microbes in the digestive systems of these creatures initiate biological decomposition through enzymatic digestion; this is accompanied by the chemical processes of oxidation, reduction, and hydrolysis, the products of which at various stages are further broken down by microorganisms. The end product of stable humus is comprised largely of the dead bodies of billions of microorganisms.

Bacteria are the most important of the microscopic decomposers. The ones that create the high temperatures of composting are *aerobic*. As the pile heats, these thermophilic bacteria predominate. As the temperature drops, mesophilic bacteria as well as fungi and actinomycetes take over. Actinomycetes may suppress bacterial populations by producing antibiot-

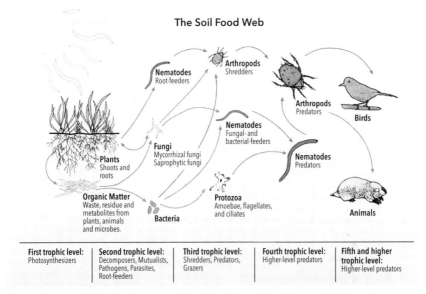

The soil community includes a vast diversity of organisms in complex interplay starting with photosynthesizers and going through decomposer organisms, parasites, predators, grazers, and higher animals whose excreta and bodies become once again part of the system. Illustration courtesy of Soil Foodweb, Inc.

ics; they tolerate dry conditions better than bacteria, and are responsible for the earthy smell of finished compost. Fungi tolerate acidity, poor aeration, and cold temperatures better than do bacteria, and may also suppress bacteria by producing antibiotics. *Anaerobic* organisms may also be present, especially inside the pile where air is limited; anaerobic composting methods have some adherents, since less nitrogen is lost to the atmosphere this way. Methane (biogas) digesters represent an anaerobic composting technology, and while their sludge can be quite valuable as a source of fertility, it should be used with caution—anaerobic organisms often cause plant diseases.

The successful composter provides the best possible conditions for the microherd to flourish and reproduce: food in the right proportions, air, warmth, and moisture. We'll discuss the more specific nutritional qualities of various compost ingredients at length a little later; the key here is always achieving a balance of several factors. Most of what the microbes need is energy, in the form of carbohydrates, composed primarily of carbon, hydrogen, and oxygen. They also need some protein, which means the addition of nitrogen-rich materials, as well as a good assortment of micro-nutrients and minerals to build enzymes. As a rule of thumb, the more diverse feedstocks it is possible to include, the more likely are the microbes to have just what they need and to produce the best quality humus.

Humus is uniform in appearance (usually dark brown) but is chemically complex and variable. Depending on the raw materials and microbial community that created it, humus can vary in nutrient content and stability. Ideally it will consist of a range of types of humus, from highly stable forms that resist further decomposition to more effective forms that are readily subject to further decomposition. The more stable forms contribute more to improving soil structure and aggregation, while the effective forms add more soluble nutrients when incorporated into the soil. A colloidal substance, humus acts as a sponge that holds air, water, and nutrients. Those nutrients that are positively charged, such as calcium, magnesium, and potassium, are made available to plants at numerous negatively charged exchange sites covering the surface of every particle of humus. A chemical buffer, humus is also able to moderate highly acid or alkaline conditions.

# three

# Temperature, Moisture, and Aeration

All composting methods work with temperature, moisture, and aeration of the mix to find the optimum relationship of each to the others. They require different regimens of monitoring conditions in the pile and adjusting them by turning, watering, or adding different materials. The quality desired in the end product will guide how time-consuming any method will be; if your goal is consistently high-quality compost finished in as little time as possible, more effort will be needed to regulate conditions in the pile. If you are not concerned with how long it takes or the consistency of the product, you can set up good conditions at the beginning and then just let it happen. If you have the space and time, why worry if it takes two years before the pile you build is ready to use?

In any case, it's aeration and moisture, together with the nutrient composition of the raw materials, that determine how hot it will get how quickly. Too much moisture reduces aeration, while too much air dries out the pile too quickly, sometimes first causing temperatures to rise too high.

## Temperature

When the pile is first constructed, it is dominated by mesophilic organisms, which begin breaking down organic matter, thus generating heat. The temperature of a good-sized pile (at least a cubic yard) will reach 140°F–160° F (60°C–70°C) in three to five days. It's a good idea to avoid exceeding 150°F (50°C), above which some of the beneficial microbes will die off. The thermophilic phase, 75°F–140°F (25°C–60°C) can last several weeks, depending on pile size and composition of ingredients. These temperatures, if maintained for at least three days, will destroy

Several factors help to heat up a reluctant compost pile, including higher concentration of nitrogenous materials (lower C:N ratio), shredding raw materials, uniform moistening of the pile, and anything that introduces air. Seasoned composters may monitor temperature by shoving an arm into the compost pile; otherwise a thermometer inserted into the pile every few days will help to monitor it. The perforated pipes sticking out of this pile reveal it to be of the static type since the pipes would interfere with easy turning.

pathogens, fly larvae, and weed seeds. Macrofauna survive this phase by moving to the cooler edges and going dormant. High temperatures of the thermophilic phase also accelerate the breakdown of proteins, fats, and complex carbohydrates like cellulose and hemicellulose.

Once the compost begins to cool it's ready for turning, which will cause the temperature to rise again owing to increased oxygen and heating of materials that were previously not decomposing. Once turning generates no more heat, the mesophilic organisms resume work as the compost *cures* or matures through the action of different microbes and chemical reactions. (Refer to pp. 41–42 for a discussion of how to assess compost maturity.)

In general, a higher concentration of nitrogenous materials (a lower carbon-to-nitrogen, or C:N, ratio) in the mix will cause temperatures to rise more quickly by stimulating microbe populations to grow faster. Shredded raw materials likewise encourage microbial activity, incorporating more air into the pile and providing more surface area for microbes to inhabit. Moistening the pile uniformly as it's created will make it heat more quickly—unless it's too wet and air is excluded. Given sufficient moisture, anything that introduces more air, such as turning, or static or forced aeration, will generate higher temperatures. Size and shape of the pile will affect how quickly it heats: larger and more rounded (as opposed to pyramid shaped) piles or windrows heat more slowly.

Regular monitoring of compost temperature will tell you whether and when it needs to be turned, or if something is preventing proper heating. Insert a 3-foot-long compost thermometer every few days to observe when the pile begins to heat, how hot it gets, and when it begins to cool. Seasoned composters may not need a thermometer at all, but test the temperature by shoving an arm into the pile.

## Moisture

A moisture content of 50–60 percent is considered the optimum range for composting. It is possible to calculate the moisture level of your mix by using charts and formulas based on moisture content of the raw materials (see the *On-Farm Composting Handbook*, listed in the "Resources" section). In general, more nitrogenous materials tend to be wetter, and more carbonaceous ones tend to be dry. Alternating layers of high nitrogen with high carbon materials permits both better microbial access to balanced nutrients and improved moisture levels. Using the squeeze test, a handful of the mixture should feel damp to the touch and as moist as a wrung-out sponge.

The best way to control moisture is to start with a good balance of wet and dry ingredients and mix them well. Adding water as the pile is built or turned is also helpful, especially if most of the materials are dry. In the Northeast's humid climate it is important to protect the pile from excess moisture, which will impede aeration and create unpleasant odors. Too much water passing through will also leach away soluble nutrients and harm water quality.

Compost-site drainage is a major concern. Runoff must be properly managed, and the potential for polluting neighboring waters is real. In addition to problems of excess moisture, a poorly drained site makes work and access difficult. Start by building the pile in a well-drained location where it will not be subject to standing water. A sloped concrete pad with a drainage system that collects the leachate is ideal, but a thick layer of wood chips or a similarly porous base works fine. Karl Hammer and Mike Merner both have leachate collection systems. Mike uses his leachate to moisten piles when they are turned, and Karl runs his leachate through a series of filters followed by a crop of Jerusalem artichokes, which then serve as forage for his free-ranging chickens.

Some kind of shelter or covering, especially if the pile is not used the same season, is important. Plastic or a tarp should only be used once a pile is finished, since it will impede aeration. A layer of hay or straw will shed water without stopping air penetration. The best compost is made under a roof of some kind that allows air circulation while protecting the pile from both precipitation and drying sun and wind. A synthetic, breathable cover developed specifically for compost, called "Compostex," is also available (see the "Resources" section). Mike Lombard uses it, and finds that it allows him to continue composting longer into the winter months when the materials would otherwise be too frozen to turn.

## Aeration

The temperature of the pile is directly related to its aeration, so normally, everything that works to increase temperature is essentially a means of increasing aeration. Shredding materials before they go into the pile and turning the pile are the most common methods of introducing air. Some farmers chop raw materials by going over them with a rotovator or using a flail chopper. A lawn mower can also be used on a smaller scale. Building the pile on a base of coarse fibrous materials, such as brush or corn stalks, or perforated pipe, introduces air without turning. Some designs create additional air passages in the body of the pile by adding more pipes or other air channels throughout it. Finally, larger scale static systems may

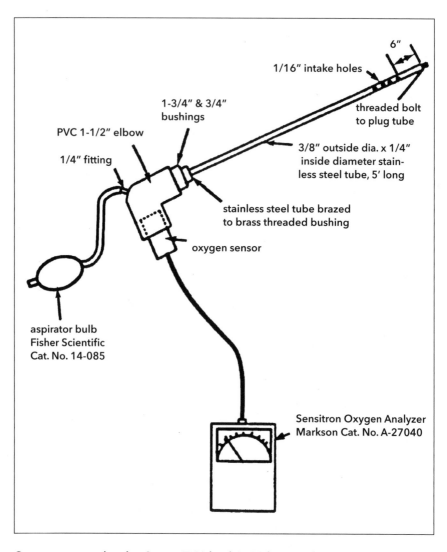

Oxygen meter and probe. *Source:* T. Richard, N. Dickson, and S. S. Rowan, *Yard Waste Management: A Planning Guide for New York State* (New York Bureau of Waste Reduction & Recycling, 1990).

employ a timed fan hooked up to underlying perforated pipes to force air through the piles, a system known as *forced aeration.*

Too much aeration can sometimes be a problem. Excess air in proportion to water can slow down the process by drying the pile, but more often, it gets too hot. Many beneficial organisms die off at temperatures of

175°F (65°C). High temperatures also volatilize more of the nitrogen and carbon in the pile, resulting in diminished nutrient and organic matter levels in the finished product. The simplest solution to excess aeration is to add water, either by hosing down the pile or by turning it and wetting each layer as it is added.

Fermentation or anaerobic composting is considered an oxymoron by many; in their view, proper humification is only achieved through aerobic processes. However, the sludge produced by in-vessel liquid composting systems, notably those used to produce methane or biogas, may be an excellent feedstock to incorporate into an aerated composting or vermicomposting system. Since only carbon and hydrogen are removed when methane is produced, this material retains more of its nitrogen than does the typical aerobic pile. This material will also tend to have a lower pH than aerobic products—think of peat moss, which is created through an anaerobic process. Anaerobic products may also exhibit phytotoxic effects, suppressing germination and actually killing plants—and so are most effectively used as a feedstock for an aerobic process.

While temperature is a pretty good indicator of the aeration status of a compost pile, precise measurement of the oxygen concentration inside the pile can give a more accurate picture of changes in microbial metabolism. The New England Small Farms Institute (NESFI) recommends using an oxygen meter, available from scientific supply houses, and probe, which can be made from parts available from a hardware store. Oxygen levels should be at least 5 percent at the beginning of composting, then decrease to nearly zero for a few days as the microbes respire. Turning or other means of introducing air will bring the level back up to 7–13 percent during active composting. If oxygen levels remain below 5 percent for more than a few days, additional air must be introduced. In static aeration systems such as the passively aerated windrow system (PAWS), oxygen levels will rise to around 21 percent once active composting stops.

# four

# Composting Methods

## High- and Low-Temperature Composting

The method you use for composting depends in large part on whether you want to use a high- or low-temperature process. Following is a summary of the pros and cons of each.

| High-Temperature Composting (temperature exceeds 131°F or 55°C for at least 3 days) | |
| --- | --- |
| PROS | CONS |
| Kills pathogens, fly larvae, and weed seeds. Produces finished compost relatively quickly. Product is well decomposed and uniform. Complies with NOP requirements. | Labor and/or equipment intensive. Releases more N through volatilization. Less plant-pathogen-suppressing qualities. |
| Low-Temperature Composting (temperature stays below 131°F or 55°C) | |
| PROS | CONS |
| Low maintenance. Conserves nitrogen. More hospitable for plant-pathogen-suppressing organisms (fungi and nematodes). | Nutrients can be lost through prolonged exposure to elements. Takes space and time before it can be used. Doesn't kill pathogens, fly larvae, or weed seeds. Finished product may be inconsistent, with undecomposed high-carbon materials that must be screened out. If contains animal manure, must be treated like raw manure per NOP. |

Once you have decided on the type of process, you can design your system along the lines of one of the methods described in the following pages. Although the descriptions of these methods involve farm-scale equipment, smaller-scale operations can follow the same basic steps—many market gardeners manage with hand tools such as manure and spading forks, shovels, and pickup trucks.

## Turned Pile or Windrow

This is the most commonly used on-farm composting system. It's basically a hot composting method, but, depending on how often the pile is turned, it may remain cool and somewhat anaerobic internally. Windrows are formed by adding alternating layers of feedstock materials, either through a manure spreader (which provides better initial aeration) or by dumping tractor bucket-loads and shaping the pile with hand tools. Where labor is more plentiful than capital for equipment, and the amount of material being composted is manageable, the whole operation can be done with manure forks. The pile is monitored and turned—the frequency depending on how quickly the finished compost is desired—using either a commercial compost turner or two tractors and a manure spreader.

The latter is a pretty straightforward method: To create the initial windrow, the first tractor is hitched to the manure spreader (PTO type, so it can unload while the tractor is stationary), while the second uses a bucket loader to transfer raw materials into the spreader. It helps to have piles of raw materials arranged along the projected length of the windrow. The full manure spreader discharges while the loader loads the right proportion of raw materials into the spreader. Once there is a pile of mixed materials behind the spreader about 4 feet high and 6 or 8 feet wide, the spreader inches forward, repeating the process (see the illustration). This method is preferable to using a bucket loader alone, since it works to aerate and mix the materials as opposed to dumping them—which can, depending on the type and condition of the materials, easily become compacted and anaerobic. The windrow is subsequently turned in the same manner, using the loader and manure spreader to move the mixture back to the original site of the raw materials. Subsequent turnings are less time-consuming because the bucket loader does not have to alternate between piles of raw materials.

Mike Merner has developed what he calls a "double-aerated" block system in which he uses a large payloader with a 5-yard bucket to form and turn the piles. Elevating the bucket to its highest point, he gently dribbles out the materials onto the pile—then repeats. In this manner he is able to build higher piles than the normal 5- to 6-foot windrow, and thus put more compost in his limited space without sacrificing quality.

Mechanized turned windrow method: The full manure spreader discharges while the loader loads the right proportion of raw materials into the spreader. When there is a pile of mixed materials about 4 feet high and 6 or 8 feet wide behind the spreader, the spreader inches forward, repeating the process.

Tom Gilbert of the Highfields Center for Composting has developed what he calls the "rolling method," using a standard tractor and bucket loader. Piles are built to be about 6 to 8 feet tall and 12 to 14 feet wide. Careful recipe development aims at a true 30:1 carbon-to-nitrogen ratio, appropriate moisture, and bulk density levels as the pile is built—Tom says it is critical to get the bulk density right in this method. The piles are monitored for temperature, moisture, and odors, and turned when the temperature drops. The pile is rolled by using the bucket to scoop up a layer two bucket-depths in from the side of the windrow, which is lifted and folded over the top to push it to the other side. In this manner the pile is slowly rolled sideways across the site as it is turned. Tom says that this system saves 50 percent of equipment and operator time compared with a complete turn. Besides aerating the areas that need it most, it also offers the advantage of avoiding unnecessary agitation. Just as in the soil, avoiding excessive disturbance promotes fungal development and minimizes loss of nitrogen and organic matter. While decomposition is slower it saves time and energy and makes a higher-quality product.

Arnie Voehringer, on the other hand, has been trained in the Lubke method of composting, which he intends to apply at White Oak Farm in Belchertown, Massachusetts. This method calls for close control and monitoring of compost conditions, with piles no higher than 4 feet and no wider than 8 feet. Carbon dioxide levels in the core are measured regularly, and if they get too high, the pile must be turned. Product quality is also evaluated, by both chemical analysis and germination testing, with the goal of producing the highest possible quality compost.

There's a variety of specialized compost turners on the market, though the cost of owning one is justified only for operations that make compost on a commercial scale. The quest for affordable composting technologies for farmers inspired Tom Gilbert of the Highfields Center for Composting to develop a design for a PTO-driven, tractor-drawn windrow turner that can be fabricated on the farm or in a local fabrication shop for approximately $3,550 in materials and 130–140 hours in labor. It can be used to manage compost windrows 5 feet high and up to 12 feet wide. The project requires welding and cutting skills, a basic understanding of hydraulics, and some basic machine-shop tools and operations. Complete plans can be downloaded free from the Highfields Center for Composting Web site (www.highfieldscomposting.org).

## Static Aerated Pile or Windrow

You don't have to turn the pile if you can introduce air, and this method most commonly uses a static pile or windrow. The pile is usually built over a series of parallel perforated drain pipes. An active-aeration system uses a fan or blower operated at intervals to draw air through the pipes by suction, which requires less power than blowing air through the system. A passive system relies on natural air convection created by temperature differences within and outside the pile, and may not aerate as thoroughly as a forced-aeration system but requires less capital investment. This method is well suited for small-scale farm operations, but unfortunately will not pass muster with the National Organic Program's rules for compost (see appendix D) unless it maintains a temperature of between 131°F–170°F (55°C–77°C) for at least three days.

The passively aerated windrow system (PAWS) was developed in Canada in the mid-1980s as a waste-management technique that saves fuel and reduces odors and insects. It entails encasing the raw compost ingredients within a shell of mature compost or peat. The windrow is formed over a base of peat or mature compost about 6 to 9 inches deep. Four-inch perforated pipes are then laid horizontally across the width of the windrow and spaced 12 to 18 inches apart, with the ends of the pipe sticking out a few inches beyond the edge of the windrow. The windrow can be formed using either a bucket loader or a manure spreader. Once the windrow is formed, an outer layer of compost or peat is then added, like the icing on a cake.

Since the pile won't be turned, it's critical to start with an optimum balance of moisture and nutrients. The best way to ensure that is to make small test piles of about 5 to 6 yards before building the windrow. After a

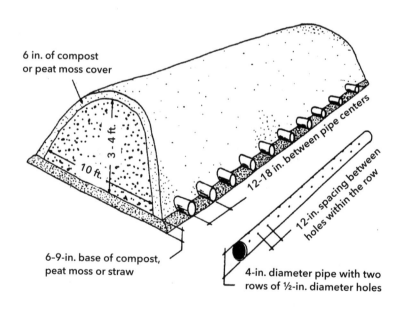

6 in. of compost or peat moss cover

3 – 4 ft.

10 ft.

12–18 in. between pipe centers

12-in. spacing between holes within the row

6-9-in. base of compost, peat moss or straw

4-in. diameter pipe with two rows of ½-in. diameter holes

Passively aerated method for composting. Adapted from Robert Rynk, ed., *The On-Farm Composting Handbook* (Northeast Agricultural Engineering Service, 1992) and the NESFI *PAWS Manual* (1993).

day or two, the test piles should heat up to at least 86°F (30°C); if heating does not occur, change the proportions of feedstock materials and try again. Although turning is not necessary, regular monitoring is important, especially when learning this technique. Temperature, oxygen, and odor should be monitored daily for the first week or two of composting, and once or twice a week after that. Some problems can be corrected by adding moisture, restricting air flow, or poking more holes in the pile (see the troubleshooting section at the end of this chapter). The compost should be stable enough to be moved to a curing area in about twelve to fourteen weeks.

The disadvantage of this system is that it's extremely difficult to correct problems of insufficient aeration, imbalanced nutrients, or excess moisture. The shell of organic materials helps to mitigate many problems, though, since users report that the peat or finished compost acts as an integrated biofilter to reduce odors by absorbing the gases produced by anaerobic decomposition, conserving water, and absorbing potential leachates. The shell also helps discourage nuisance insects by eliminating access to their food and by remaining relatively cool.

Arnie Voehringer worked with the PAWS system at the New England Small Farms Institute, using dead fish from an aquaculture operation and rabbit manure. He says that it worked well, produced good-quality compost, and did not smell. He does say that maintaining the right moisture level is critical to success. Mike Merner says he has tried using a PAWS system and found it to be a big headache when trying to remove the finished compost with his loader without mangling the aeration pipes.

## Vermicomposting

Technically, vermicomposting is more akin to livestock production, since the process relies not on microorganisms, but on worms that digest raw organic materials and excrete them as nutrient-rich castings. Earthworms and microorganisms join forces in a bio-oxidation and stabilization process that does not have a thermophilic stage. The earthworms are the agents of turning, fragmentation, and aeration.

The most common earthworms used for vermicomposting are brandling worms (*Eisenia foetida*) and red worms or red wigglers (*Lumbricus rubellus*).

Often found in aged manure piles, red worms generally have alternating red and buff-colored stripes. They are not to be confused with common garden or field earthworms (*Allolobophora caliginosa* and other species), which, while larger and more desirable as bait species, will migrate out of the beds. Worms can be ordered from a variety of dealers.

Many municipalities use vermicomposting systems to turn their wastes into a soil amendment and simultaneously produce crops of worms that go to make high-protein fish or poultry feed or are sold as bait. Cuba is a large producer of worm compost, utilizing both animal manure from the cattle industry and mountains of filter cake, a waste product of the sugar-processing industry. Worm beds can be very small (I had one in my urban apartment for two years) to industrial-scale continuous flow reactors where raw organic matter is fed in at one end and worms and their castings are removed at the other.

Farm-scale vermicomposting systems generally utilize beds or windrows on the ground containing material up to 18 inches deep. Prepare the site as for any compost method, with good drainage and shelter from the weather. Shredded paper or cardboard make good bedding, as do shredded leaves, straw, hay, or dead plants, sawdust, peat moss, or aged manure. Fresh manure or other high nitrogen materials that will heat up quickly should not be used, since the temperatures of thermophilic composting will kill or drive away the worms. The same principles of balanced nutrient sources, aeration, and moisture apply as for hot composting. The bedding material is thoroughly moistened—about the consistency of a damp sponge—before adding the worms. Some soil or sand can be added to help provide grit for the worms' digestive systems.

Worms, like other compost organisms, require a good balanced feedstock, appropriate pH, warmth, aeration, and the right moisture level to allow the animals to breathe through their skins. Under optimum conditions, red worms can eat their own weight in organic matter in one day. On the average, however, it takes approximately 2 pounds of earthworms (approximately 2,000 breeders) to recycle a pound of organic waste in twenty-four hours. The same quantity of worms requires about 4 cubic feet of bed to process the organic material (1 cubic foot of worm bed per 500 worms). Add worms to the top of the moist bedding when they arrive. The worms will disappear into the bedding within a few minutes.

Red worms can survive a wide range of temperatures (40°F–80°F), but they reproduce and process food waste at an optimum bedding temperature range of 55°F–77°F. Worms will require some form of insulation to survive the Northeast winter, such as a thick layer of straw, sawdust, or shredded paper laid over the bed before temperatures drop below freezing.

There are three basic ways to separate the worms from the finished compost. One way involves moving the finished compost and worms over to one side of the pile or bin and adding new bedding material and organic waste to the other side. Worms in the finished compost should move over to the new bedding with the fresh food waste. The finished compost can then be removed. A second way to remove the worms is to build a harvester frame of 2 × 4s with a ⅜₆-inch mesh bottom. Place the worm compost on the frame and sift the worms out. Larger pieces of compost can be returned to a new batch of bedding and worms.

As a final alternative, the compost also can be placed in small piles on a tarp in the sun. Because worms don't like light, they will wiggle to the bottoms of the piles. After waiting ten minutes, remove the upper inch or more of finished compost from each pile until you run into the worms. Allow the worms to again wiggle to the bottom of the pile and repeat the process, sometimes known as *scalping*. Combine what's left of the small piles into one big pile and again repeat the process. You should eventually end up with a pile of finished compost and a ball of worms. The worms can be used to start a new bed, fed to poultry or fish, or sold to other farmers.

Some research suggests that vermicomposting is comparable to hot composting in reducing populations of pathogenic microorganisms. Vermicomposting may also be better than hot composting in reducing the bioavailability of heavy metals, and there is evidence that the final product may contain hormone-like compounds that accelerate plant growth. An experiment at the Soil Ecology Laboratory of Ohio State University studied a continuous vermicomposting process for different mixtures of pig-manure slurries and agroforestry by-products and compared them with controls produced through a hot composting process. Results showed higher nitrification in the earthworm compost, as well as a decrease in the carbon from fulvic acids and an increase in the percentage of the carbon from humic acids, indicating that nitrogen was more readily available.

Germination tests showed that the phytotoxicity of the original mix was eliminated by the vermicomposting process, which supports the argument that vermicomposting creates as stable a finished product as do hot composting methods.

Because vermicomposting occurs at low temperatures, at this time the National Organic Program (NOP) does not consider the product of vermicomposting to be "compost" under its rules. However, in a recent draft guidance document entitled "compost and vermicompost in organic crop production" the NOP has indicated that vermicomposting is an acceptable composting method, and offers information about what methods should be used. See chapter 9 for an explanation of the NOP's composting rules and how they came about.

## Diagnosing Problems

What happens when the compost pile doesn't work according to plan? The principles of making good compost may be simple, but the practice takes, well, practice. Expert composters agree that composting is more akin to an art than an exact science—rules, recipes, and formulas can only be considered guidelines. As Karl Hammer said, "I've never seen two turds that were exactly alike." The guidelines in the table can help you figure out what happened and how to fix it.

| What Went Wrong? | | |
|---|---|---|
| Conditions | Possible causes | How to fix it |
| *Initial stages–within the first week or so:* | | |
| Failure to heat | C:N ratio too high, or pile too dense or wet (not enough air), or pile too small (not enough critical mass) or too big (not enough air). | Turn and add more high-N materials. If static system, poke more holes. Make the pile closer to the optimum size (no more than 5 feet high). |
| Too much heat | C:N ratio is too low, or there is too little moisture. | Turn, add more high-C materials, and moisten while turning. Spray pile, or if static pile poke holes and inject some water. |
| Foul odors | Too little air or too much moisture (anaerobic conditions). C:N ratio is too low. | If pile is heating properly, wait a day or two to see if smell stops. Turn and add more dry high-C materials; if static, add air passages, consider rebuilding pile, adjusting mix. |
| *After turning, or when compost should be ready to use:* | | |
| Center is dry, materials not broken down | Too little moisture; materials not well enough mixed. | Turn and moisten as the pile is built. Cover with plastic or tarp to prevent drying out. |
| Large undecomposed chunks | Materials of high lignin content (e.g., wood wastes) were used; nonbiodegradables (e.g., plastic, metal) present in materials. | Screen out chunks and use as base of next pile. Ensure that raw materials come from well-separated source(s). Use a shredder or grinder for very coarse raw materials. |
| Poor compost quality (low nutrients, phytotoxicity) | Compost is not yet finished, still unstable. Raw materials lacked balanced nutrients. | Allow more curing time (move curing area under cover). Use more diverse raw-materials sources and/or use mineral amendments when building pile. |

# five

# Materials

The decision about whether and how to compost often hinges on the availability of good feedstock—appropriate organic waste materials. (Since the NOP doesn't require compost feedstock to be certified organic, the term "organic materials," as used here, means *materials based on living things*, as opposed to inorganic ones). Dairy farmers have a ready source of manure, but unless they use a lot of bedding, may have difficulty finding enough dry, carbonaceous material to mix with it. Farms close to cities may have abundant access to yard trimmings and leaves, as well as food scraps from supermarkets and restaurants. Food processing wastes may be problematic because of high moisture content. Animal carcasses and slaughterhouse offal pose public health problems if not handled properly—the Highfields Center for composting teaches farmers how to safely compost animal mortalities. Instructional videos are also available from Cornell or the Vermont Association of Natural Resource Conservation Districts (see the "Resources" section). Concerns about using compost from wastes that have been heavily treated with pesticides, and now the possibility of GMOs (genetically modified organisms), cause uncertainty about accepting materials that might otherwise be ideal compost ingredients. If you intend to sell finished compost, a consistent supply of feedstock ingredients is important to producing a reliably uniform product.

As solid-waste disposal becomes costlier, potential sources are expanding. Urban and rural waste-management districts frequently make available lists of industrial sources of organic wastes, and many of them will transport the materials to your location and pay a fee for the service of disposing of them. Mike Merner, who uses about 10–20,000 cubic yards of raw materials a year, charges tipping fees for hard-to-dispose-of wastes such as fish gurry, but other organic wastes, such as farm and zoo manures and seaweed from local beach clubs, are delivered to his site free of charge. Mike Lombard, who sells about 2,500 cubic yards of compost and soil blends per year in southern New Hampshire, keeps it simple by using only well-bedded horse manure mixed with cow or chicken manure.

The Highfields Center for composting has initiated a program called "Close the Loop Vermont" to collect food wastes from cooperating towns and waste-management districts and work with local farmers to produce compost from the food scraps combined with dairy manure. Initiated through grant funds, but transitioning to a fee-for-service model, the program includes public outreach and education, especially in local schools, to recruit waste generators and train them to keep the waste as trash free as possible. The goal of Close the Loop Vermont is to develop the infrastructure to meet all the state's food-waste composting needs, and simultaneously generate in-state sources of fertility.

If regular deliveries of organic wastes are in your plan, you will have to pay attention to regulations concerning waste-disposal facilities (see chapter 9). If a particular industry's organic waste product becomes available, it's always advisable to test samples before embarking on a plan to compost it. The "Compost Raw Materials" table at the end of this chapter provides some information about approximate C:N ratios, nitrogen content, moisture level, and possible concerns for materials that are commonly available in the Northeast. If the material that you have available is not specified there, look for something that is similar—the figures should be considered rough estimates that allow you to judge the range of proportions needed for your situation.

The ideal C:N ratio of 20–30:1 can be created either by starting with fairly well-balanced raw ingredients or by forming the pile with alternating layers of high carbon and high nitrogen ingredients. Keep in mind that the ratio is based on weight. Highly carbonaceous materials, such as straw, dry leaves and sawdust, tend to be dry, light, and bulky, while highly nitrogenous materials such as fish wastes, poultry manure, and green yard trimmings tend to be wet, heavy, and dense. If you visualize 10 pounds of poultry manure next to 250 pounds of sawdust, you will get a good sense of the relative sizes of the piles of ingredients needed to achieve the best proportions.

Besides C:N ratio, it is important to pay attention to the moisture content and density of the feedstock materials. Sophisticated calculators are available to determine ideal recipes, based on nutrient balance and moisture content of the intended ingredients (see the table at the end of this chapter). For any material, these values vary, depending on a range of conditions, so that accurate assessment is difficult without having it tested by a laboratory. Smaller operations that rely on one or two primary ingredients can judge more easily. Over time, you learn to judge the moisture balance that works best by feel, and to eyeball a pile of raw ingredients and decide how much of what is needed to balance the carbon and nitrogen.

## Compost Additives and Inoculants

You may decide to augment the nutritional quality of your compost by adding minerals, whether liquids or powders. If you do it at the mixing stage, most nutrients, especially phosphate and micronutrients, will be converted into a more bioavailable form through the composting process. Therefore it makes great sense to apply rock powders such as colloidal phosphate and greensand via compost rather than spreading them directly in the field. Adding micronutrients to compost can ensure balanced trace-mineral nutrition and help avoid problems of spot imbalances that can happen when these materials are spread directly. If certain nutrients, such as boron, tend to be deficient in your area, this is an excellent way to provide them. (Test the soil before doing this to be sure the nutrient is needed on your farm.) Many commercial compost products contain added minerals. Some companies blend concentrated organic wastes such as blood meal into their finished product as a means to ensure a consistent nutrient analysis, or to boost the analysis to higher levels.

Most research results show little or no benefit from any of the various compost inoculants on the market. A small amount of healthy topsoil will serve to introduce local microorganisms into a batch of finished compost. (See "Effective/Indigenous Microorganisms" in chapter 8.) Many farmers, both Biodynamic and non-Biodynamic, also use Biodynamic compost preparations.

## Biodynamic Preparations

The Biodynamic method starts with a windrow—static or turned—into which special compost preparations are immediately introduced. Although these preparations are rich in microorganisms and trace minerals, their purpose is not to inoculate the compost but to activate and enhance the etheric or life forces on the farm. Five of the preparations (numbers 502–506) are inserted into holes made in the windrow, so that each material is placed about 5 to 7 feet from the other, about halfway up the side of the windrow. Long windrows get a series of preparations added in repeating sequence. A small amount is placed in each hole, then covered with manure. The whole pile is then watered with a preparation made from valerian flowers. This is done only once, and does not have to be repeated when the pile is turned. Cow manure is often considered to be an essential ingredient in a Biodynamic compost pile, since the particular microbial contribution of the cow's digestive system is considered integral to creating quality humus.

The preparations themselves consist of combinations of plant and animal substances, prepared according to specific directions provided by Rudolf

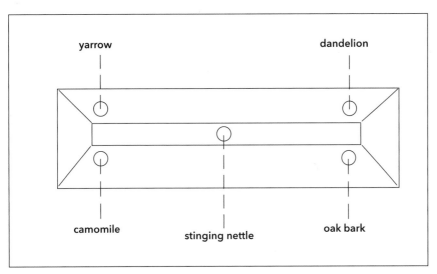

Example for the placing of Biodynamic preparations in a compost heap. From H. H. Koepf, B. D. Pettersson, and W. Schaumann, *Bio-Dynamic Agriculture: An Introduction* (Spring Valley, NY: Anthroposophic Press, 1976).

Biodynamic preparations include oak bark, dandelion, valerian, cow organs, etc. Cow manure is often an essential ingredient in a Biodynamic compost pile. Cycles of the moon are symbolized in the background.

Steiner in his agriculture lectures. Plants used include yarrow blossoms, chamomile blossoms, stinging nettle, oak bark, dandelion flowers, and valerian flowers. Animal organs used include a bovine mesentery, bovine intestine, any domestic animal skull, and a stag bladder. The process of making these preparations is intricate and lengthy, generally involving allowing the original materials to decompose over the course of several months or a year. Fortunately, you can buy the preparations ready-made, as well as BD Compost Starter, originally developed by Dr. Ehrenfried Pfeiffer, which is simpler to apply than the individual preparations. It consists of all the BD compost preparations as well as BD No. 500 and a mixed culture of microorganisms, and is available from the Josephine Porter Institute (see the listing in the "Resources" section).

Biodynamic compost, according to research cited by Koepf et al., shows higher cation exchange capacity than compost made in the same place using the same raw materials without Biodynamic preparations. According to Nicholas Lampkin, the positive effects of using Biodynamic preparations have been established through field, pot, and laboratory experiments. "These results include morphological and physiological changes in contents, as well as other factors such as protein balance, enzyme activities, keeping quality, and taste of produce" (Lampkin, 654). Research at Washington State University has shown a significant effect on compost treated with BD preparations, including higher temperatures, faster maturation, and higher nitrate levels (Diver, 5).

## Materials to Avoid

It should be obvious that inorganic materials—anything that is not of plant or animal origin, except for mined mineral amendments—should not be used in compost. Certainly the National Organic Program permits only naturally mined (nonsynthetic) fertilizer materials, such as ground rock powders. Use of any synthetic micronutrient sources also requires documentation that the nutrient is deficient in your soil or crops. This means that synthetic nitrogen fertilizers, sometimes recommended to activate compost mix that's extremely high in carbon, aren't to be used. Other materials, even though they may be by-products from processing animal or plant materials, are technically considered to be synthetic industrial wastes. An example of this is leather waste, which is laden with heavy metal (chromium) and considered synthetic under the NOP definition. By this definition municipal sewage sludge or biosolids would also be considered to be synthetic, and therefore would not be permitted for organic production—even without the explicit prohibition included in the NOP final rule.

One of the biggest headaches for composters who collect waste from municipal sources or even dairy farms is the presence of nonbiodegradable material such as plastics and metal. Karl Hammer has worked with his sources to educate them on the importance of source separation, but says he still has problems with things like syringes tossed into the drop by vets. Karl uses his 500 laying hens to pull the plastic out of food waste he receives from the New England Culinary Institute, which he feeds to the chickens before it is composted. Although plastic can be ground up and camouflaged, or screened out of the finished product, Karl believes it has a deleterious effect on compost quality, and there's no telling what kinds of toxic compounds are released when it breaks down. Mike Merner has written agreements with his sources specifying that the product must be clean and source-separated.

There are few pesticides and herbicides now permitted on crops in the Northeast that might not completely break down in the composting process. Some of the newer herbicides, such as clopyralid, picloram and triclopyr, have been found to break down very slowly in compost and have caused serious phytotoxicity problems in the Northwest. These herbicides may also be found on grass trimmings in the Northeast, although not at the levels found in Washington State. When in doubt, you should get a laboratory analysis of residues in the raw waste product. In addition, you could run a trial batch of compost with the ingredient in question and then have the end product analyzed. The problem of how to handle organic wastes with high heavy metal content is a difficult one. Composting will do nothing to decrease the quantity of heavy metal in the product, rather it will concentrate the metal as the bulky waste is reduced. However, because of its chelating properties, compost will bind the metals in unavailable form. While this is helpful when there are small amounts of these toxic substances present, deliberately adding them to your soil system invites the possibility of releasing them as conditions change and the humus is subject to mineralization later on. The table at the end of this chapter notes materials that should be evaluated for contaminants such as pesticide residues or heavy metals.

Another extremely difficult issue is that of organic wastes derived from genetically modified organisms (GMOs). Corn, soybeans, canola, and potatoes now could all possibly be GMOs, which are not permitted anywhere in organic production under the NOP. Research is still inconclusive as to how completely the composting process breaks down the genetically altered DNA, and there are no definitive limits (at least in the United States) for GMO contamination in an organically grown product. In the case of wastes of crops that may be GMOs, (e.g., spoiled corn silage), it is best to make sure that the material was not grown from genetically modified seed. There is, however, no clear evidence that GMOs in the feedstocks will remain in the finished compost. Karl Hammer, who uses spoiled corn silage in his composting operation, argues that it is better to compost these materials than to let them decompose in the environment. The new Canadian Organic Regime (COR) includes a note in its regulation on sources of manure that organic operations should avoid manure from livestock operations using GMOs or their derivates in animal feeds (CAN/CGSB-32.310-2006—Amendment No. 1, December 2006, Section 5.5.1).

A list of recipe calculators—and potting-soil recipes—can be found in the appendix (A and C).

| Compost Raw Materials | | | | | |
|---|---|---|---|---|---|
| Material | C:N ratio | % N dry wt. | Moisture content % wet wt. | Bulk density lbs./cu.yd. | Sources, comments, and cautions |
| Crop residues, fruit/vegetable processing wastes | | | | | |
| Apple filter cake | 13 | 1.2 | 60 | 1,197 | Orchards, cider mills, food processors |
| Apple pomace | 48 | 1.1 | 88 | 1,559 | Orchards, cider mills, food processors |
| Coffee grounds | 20 | 2.0 | –* | – | Restaurants, coffee shops |
| Corn cobs | 98 | 0.6 | 15 | – | If off-farm, note GMO status |
| Corn stalks | 60–73 | 0.6–0.8 | 12 | 32 | If off-farm, note GMO status |
| Cranberry filter cake | 31 | 2.8 | 50 | 1,021 | Juice processors |
| Cranberry plants | 61 | 0.9 | 61 | – | Cranberry farms |

| Material | C:N ratio | % N dry wt. | Moisture content % wet wt. | Bulk density lbs./cu.yd. | Sources, comments, and cautions |
|---|---|---|---|---|---|
| Cull potatoes | 18 | 0.35 | 78 | 1,540 | If off-farm, note GMO status |
| Fruit wastes | 20–49 | 0.9–2.6 | 62–88 | – | Food processors |
| Potato process sludge | 28 | – | 75 | 1,570 | Note GMO status |
| Potato tops | 25 | 1.5 | – | – | If off-farm, note GMO status |
| Rice hulls | 113–1120 | 0–0.4 | 7–12 | 185–219 | Mills, cereal manufacturers |
| Soybean meal | 4–6 | 7.2–7.6 | – | – | Note GMO status |
| Tomato process waste | 11 | 4.5 | 62 | – | Food processors |
| Vegetable wastes | 11–13 | 2.5–4 | – | – | Food processors, packers |
| Fish & meat processing | | | | | |
| Blood waste (slaughterhouse waste, dried blood) | 3–3.5 | 13–14 | 10–78 | – | Slaughterhouses |
| Crab, lobster wastes | 4.0–5.4 | 4.6–8.2 | 35–61 | 240 | Shellfish processors; contains chitin, high Ca |
| Fish process sludge | 5.2 | 6.8 | 94 | – | Fish processors |
| Fish wastes (gurry, racks, etc.) | 2.6–5.0 | 6.5–14.2 | 50–81 | – | Fish processors |
| Mixed slaughter-house waste | 2–4 | 7–10 | – | – | Slaughterhouses |
| Poultry carcasses | 5 | 2.4 | 65 | | Broiler operations |
| Paunch manure | 20–30 | 1.8 | 80–85 | 1,460 | Slaughterhouses; partially digested feed–possible GMO issues. |
| Manures | | | | | |
| Broiler litter | 12–15 | 1.6–3.9 | 22–46 | 756–1,026 | Type and amount of bedding determines values for any live-stock manure. |

| Material | C:N ratio | % N dry wt. | Moisture content % wet wt. | Bulk density lbs./cu.yd. | Sources, comments, and cautions |
|---|---|---|---|---|---|
| Cattle | 11–30 | 1.5–4.2 | 67–87 | 1,323–1,674 | |
| Dairy | 13–18 | 2.7–3.7 | 79–83 | – | |
| Horse | 22–50 | 1.4–2.3 | 59–79 | 1,215–1,620 | |
| Laying hens | 3–10 | 4–10 | 62–75 | 1,377–1,620 | |
| Sheep | 13–20 | 1.3–3.9 | 60–75 | – | |
| Spent mushroom | 16.9 | 1.8 | – | – | High levels (>2%) P, K compost |
| Swine | 9–19 | 1.9–4.3 | 65–91 | – | |
| Turkey litter | 16 | 2.6 | 26 | 783 | |
| Straw, hay, silage | | | | | |
| Corn silage | 38–43 | 1.2–1.4 | 65–68 | – | Note GMO status |
| Hay-general | 15–32 | 0.7–3.6 | 8–10 | – | Range reflects legume content as well as maturity of hay. |
| Oat straw | 48–98 | 0.6–1.1 | – | – | |
| Wheat straw | 100–150 | 0.3–0.5 | – | – | |
| Wood and paper | | | | | |
| Bark-hardwoods | 116–436 | 0.10 | – | – | Range reflects differences in species as well as how dry it is. Good substrate for disease-suppressing organisms. |
| Bark-softwoods | 131–1,285 | 0.04–0.39 | – | – | |
| Corrugated cardboard | 563 | 0.10 | 8 | 259 | Needs to be shredded. |
| Lumbermill waste | 170 | .13 | – | – | |
| Newsprint | 398–852 | 0.06–0.14 | 3–8 | 195–242 | Most inks currently in use are safe to compost. |
| Paper fiber sludge | 250 | – | 66 | 1140 | Check for possible dioxin contamination. |

| Material | C:N ratio | % N dry wt. | Moisture content % wet wt. | Bulk density lbs./cu.yd. | Sources, comments, and cautions |
|---|---|---|---|---|---|
| Paper pulp and mill sludge | 54–90 | 0.56–0.59 | 81–82 | 1,403 | |
| Sawdust | 200–750 | 0.06–0.8 | 19–65 | 350–450 | From untreated wood only. |
| Telephone books | 772 | 0.7 | 6 | 250 | |
| Wood chips | – | – | – | 445–620 | From untreated wood only. |
| Hardwood chips, shavings, etc. | 451–819 | 0.06–0.11 | – | – | |
| Softwood chips, shavings, etc. | 212–1,313 | 0.04–0.23 | – | – | From untreated wood only. |
| Yard wastes and other vegetation | | | | | |
| Grass clippings | 9–25 | 2.0–6.0 | – | 300–800 | Check for herbicide residues; range reflects loose vs. compacted clippings. |
| Leaves | 40–80 | 0.5–1.3 | 38 | 100–500 | Range reflects loose and dry vs. compacted and moist. |
| Seaweed | 5–27 | 1.2–3.0 | 53 | – | |
| Shrub trimmings | 53 | 1.0 | 15 | 429 | |
| Tree trimmings | 16 | 3.1 | 70 | 1,296 | |
| Water hyacinth-fresh | 20–30 | – | 93 | 405 | Aquatic weed disposal. |

Source: Adapted from Rynk, ed., *On-Farm Composting Handbook*.
*– = Data not available from sources used.

# six

# About Costs

If you plan to use composting as the basis for an income-generating enterprise, be prepared to give the same time and attention to business planning and management as you would any other farm-based enterprise, whether it be milk, meat, fruits, or vegetables. (See Elizabeth Henderson and Karl North, *Whole-Farm Planning,* 2004, another handbook in this NOFA series). Considerations such as waste-disposal regulations, marketing, and control of product quality will need to be included in a decision to turn compost making into an income-generating enterprise. As is often said, the compost maker is a farmer of microbes, no less a farmer than a dairy farmer or a vegetable grower.

The primary expenses involved in composting are those of site preparation, trucking materials, equipment to pile, aerate, and spread the compost, and labor. Very small operations need almost nothing in the way of specialized equipment—depending on the type of system used, a few hours of work with a manure fork and shovel every few weeks may be the only cost. In fact, some farmers have found that they can recapture equipment and labor costs by "selling" waste disposal for local food processors or other producers of organic wastes, as well as selling the finished compost.

## Composting Budgets

What follows are guidelines for projecting costs and income for a farm-scale operation making compost for on-farm use. All the figures should be considered ballpark estimates, based on published studies. Actual costs can differ considerably from situation to situation. Estimates given here of time needed for various operations are more likely to accurately reflect actual experience on the farm. The figure that comes out the other end is the critical one of cost per cubic yard or per ton of finished compost;

this is a key indicator of whether you are better off making your own or buying it elsewhere. An example of a composting budget sheet is provided in appendix B. (*Note:* Unless otherwise indicated, the costs and price estimates given here are 2003 figures.)

## Equipment Needs

Small-scale producers can make compost by hand, but a larger dairy or field-crop operation requires mechanization. The necessary equipment may already be on hand, or can often be acquired used. Custom compost turning services exist, and using them eliminates the need to buy and maintain equipment for that part of the job. Here are estimates of the cost of equipment and inputs other than materials for establishing a composting operation and a list of which tools or inputs apply to which method.

| Tools, Equipment, and Materials for Commercial Composting | | | | |
|---|---|---|---|---|
| Tool/input | Capital cost and/or per hr. | Turned windrow | Static aerated windrow | Vermicompost |
| Shredder or flail chopper | Flail $8–$10K; shredder/ screener $150K | # | # | # |
| Compost turner | Smallest PTO-driven turner ~$10–$15K; self-powered $115K; custom turning $100/hr. | # | X | X |
| Tractor with loader | $15–$50K; $15–$30/hr. | + | + | + |
| Manure spreader | $500 and up | + | # | # |
| Shelter or cover | Compostex compost covers: $300–$600/roll | # | # | + |
| Monitoring tools | $100–$200 (thermometer and oxygen probe) | # | # | # |
| 4" perforated drain pipe | $70/100 ft. | X | + | X |
| Screener | $50K | # | # | # |
| Worms | $15–$20/lb. incl. shipping | X | X | + |
| Key: + Necessary or recommended X Not applicable # Optional | | | | |

## Site Preparation

Composting requires dedication of a certain amount of land with the same characteristics as prime cropland: excellent drainage, access for equipment, and minimal slope. Site preparation may entail grading and laying gravel or pouring a concrete pad. An intensively managed turned-windrow setup requires space for equipment to move around; a static aerated or cold compost system doesn't require as much room around a windrow. Depending on the technology used, an acre of land can process from two or three thousand cubic yards to several tens of thousands of cubic yards of compostables per year. A separate curing area, under shelter, is also advisable.

## Cost of Materials

In most cases the cost of materials themselves will be minimal, reflecting primarily loading and trucking expenses. In some cases acceptance of organic waste materials will represent a source of revenue through tipping fees. Depending on the area and the type of materials, waste-disposal revenues can range from zero—free delivery of materials—to $25 per ton or more. Sometimes the purchase of particular materials may be justified by compost-quality considerations, and the desire to use special inoculants or mineral amendments. Karl Hammer has made a decision to pay $3 per cubic yard for good quality, clean manure as opposed to charging tipping fees, since he feels the leverage it gives him on quality saves him time and money and produces a superior product. In other cases, a composter will elect to buy dry high-carbon material, such as sawdust, to balance a readily available source of wet high-nitrogen material, such as fish wastes. Depending on the availability of labor and the method chosen, it may make sense to invest in an airy carbonaceous material such as straw to provide enough aeration to eliminate the need to manipulate the pile. One case study estimates the cost of raw materials, including trucking, at $5 per ton.

## Labor and Time

In general, if you choose to minimize capital costs, you'll raise your expenses for labor—and vice versa. If you already own some farm equipment, it makes sense to use what you have, but sometimes an upgrade of equipment that can serve more than one function is warranted. Essentially, you'll need to decide if you can handle the labor requirements for the amount of compost you want to make, perhaps fitting it around other demands for labor and equipment. The following information comes from various studies that looked at time requirements for composting on various scales with different methods.

Loaders on farm tractors typically have about a 1-cubic-yard capacity (which translates as 900–1,500 pounds, depending on moisture content), while larger-scale municipal composting operations may use equipment with a 2- to 5-cubic-yard bucket. One study found that a 40-horsepower tractor and ⅓-cubic-yard bucket could turn about 20 cubic yards of material per hour. With a 1-cubic-yard bucket, that means it would take over three hours to turn 200 cubic yards. Steve Wisbaum estimates that using a tractor and manure spreader, you can turn about 40–80 cubic yards per hour, while his custom compost turner can turn 300–600 cubic yards per hour.

The differences between cold composting and static pile systems (whether passively or actively aerated) in terms of labor needs can be significant—in addition to saving wear and tear on equipment. Based on the figures given above, 200 cubic yards of compost turned five times to comply with the NOP requirements would need about 15 hours of time by two people, two tractors, and a bucket loader. (See chapter 9 for more on the NOP requirements.)

## Value of Product and Potential Income

A study done in 1990 in Pennsylvania found the prices of bulk compost to vary considerably (from about $10 to over $300 per dry ton), and this is doubtless true today as well. The higher figures reflect products supplemented with minerals to boost the guaranteed analysis and distributed in bags. In 2003, Mike Merner charged $40/cubic yard for his finished compost in bulk, while Karl Hammer charges $33 and Mike Lombard $30.

Prices in 2010 for professionally made compost in Vermont ranged from $40 to $65 per cubic yard. Highfields Center for Composting currently charges $46 for a single cubic yard. All suppliers charge less for larger quantities, while farm compost ranges from $20 to $45 per yard.

## Overall Costs of Compost Production

The financial variables involved in running equipment can be complex. Depending on the scale of the operation and the technology used, initial costs of site preparation, planning, and permitting may run from a few hundred to thousands of dollars. Municipal compost operations report costs ranging from a few dollars per ton to more than $100/ton. A study done by Cornell in 1993 evaluated the literature on costs of composting. It provides detailed analysis of labor, equipment, and materials costs for several New York farms of different scales. Results are summarized in a table showing time and costs of making compost by turning windrows four times annually, with a variety of different equipment scenarios and three different volumes (1,000, 5,000 and 15,000 cubic yards). At the lower volume, the cheapest method was a tractor and loader (85-horse-power tractor and $6,000 loader) at $1.12/cubic yard. At the 5,000-cubic-yard volume, the cheapest method was the small, PTO-driven windrow turner with a 40-horsepower tractor, which came out at $0.58/cubic yard. This system was also the winner in the largest volume, at $0.28/cubic yard. Hourly operating costs for the various systems ranged from $10 to $22.

The cost of intensively turned hot composting can be significantly higher than for cold compost that is allowed to sit for a longer time before it is mature (see chapter 4 on other pros and cons of hot versus cold composting). Care in site preparation and establishing initial materials mixes with good aeration can pay off in producing high-quality compost with minimal manipulation costs. A few studies have been done that examine the cost and quality of end products with intensive composting. One series that compared the time efficiency for creating mature or stable compost (determined by no increase in pile temperature) showed that frequent turning (twice weekly) required 106 days versus 123 days for no-turned piles for dairy manure, and 130 days versus 145 days for poultry manure. A comparison showed the cost per wet ton to be $3.05 for the no-turned pile versus $41.23 for the compost turned twice weekly.

Steve Wisbaum's case studies come up with a total cost per cubic yard of finished compost that ranges from a little more than $1 at Shelburne Farms, where they compost their own dairy manure and make over a thousand cubic yards of compost, to almost $12/cubic yard at Long Wind Farm, where they buy in all their materials and make about 500 cubic yards of compost. Cornell's Co-Composter program performed a study comparing actual costs of composting on four farms versus costs projected using the Co-Composter model, and came up with actual costs in the vicinity of $18/cubic yard. (See appendix A.) The professional composters I interviewed were reluctant to hazard guesses as to their actual costs per yard, although Mike Merner figures that, considering operating costs as well as overhead, his compost costs him around $20–$25 a yard to make.

Tom Gilbert suggests that the numbers for Shelburne Farms given above do not account for purchasing carbon materials, which is important to producing high-quality compost. He believes that the toughest aspect of cost assessment is creating a consistent baseline for cost of investments and management. Tom estimates that making high-quality compost using an on-farm manure source, combined with an equivalent carbon input, runs at least $17–$20 per yard. "This assumes you fully value the cost of running a loader and paying someone a living wage," he adds. Loader costs, including fuel, grease, maintenance, and repair may run $35–$65 per hour depending on size and model—a common mistake is failure to adequately value time and overhead.

# seven

# When Is Compost Finished? And How Is It Applied?

## When Is It Finished?

Depending on how often it has been turned or aerated, and on the air temperature and fineness of the original materials, a batch of compost can be finished in as little as four weeks. Its total mass will be reduced by about half, and it won't heat up again when it's turned. The original materials will be indistinguishable, the color and texture will be relatively uniform, and the odor will be pleasant and earthy. Bits of coarse carbonaceous materials like wood chips or corn cobs may still remain, and these can either be screened out or left in the mix to decompose in the soil, depending on the intended use.

There are several methods for assessing whether compost is fully mature and stable, something that's critical for applications such as seed starting. One way is to use a sample to conduct a seed-germination test. Unfinished or anaerobic compost commonly releases ammonia and other phytotoxic compounds, which will inhibit germination and root development. To do a germination test, use two equal amounts (twenty or more) of seeds—cucurbit family seeds and cress are particularly sensitive to immature compost. Plant the first batch in the compost sample. The second should be germinated in a moist paper towel or similar non–nutrient medium, and both should be kept in the same conditions of darkness, warmth, and moisture. If germination is significantly lower in the compost sample, and/or seedlings less robust, it is not stable enough.

A more sophisticated test is to send the sample to a lab to measure the ratio of nitrate to ammonium forms of nitrogen. Fully mature compost will contain several times more nitrate than ammonium forms. High-quality finished compost will show

- ammonium at less than 2 mg/kg;
- nitrite that is not detected;
- nitrate that is less than 100 mg/kg in cold months and less than 300 mg/kg in warmer months;
- a pH of less than 8;
- negative test for sulfides; and
- a germination index greater than 90 percent.

Finally, Woods End Laboratory has developed the Solvita® test kit for compost maturity. This employs a specific volume of sample that is moisture adjusted and put into a sealed jar for up to seventy-two hours. Two paddles—one that tests for carbon dioxide and one for ammonia—are then placed in the jar for four hours, and the color is compared to a chart. A value of 7–8 on the carbon dioxide paddle indicates that the compost is very mature, while a value of less than 5 indicates immature compost. Ammonia levels may range from low to high on a scale of 1–5.

## Compost Quality

The question of how to evaluate compost quality depends largely on how the compost will be used. If it is all for the farm, to be spread or incorporated into the soil to build fertility and provide nutrients to crops, it is less critical to monitor quality than if it is being sold or used for specific applications such as greenhouse potting mixes or for disease suppression (see the "Special Uses" section later in this chapter). Only recently has the need to establish regulatory standards emerged, largely as a result of widespread municipal composting operations. The products of these operations have raised concerns about residues of hazardous substances such as heavy metals, toxic chemicals (e.g., dioxin), and pathogens. More recently still, private-label schemes have emerged that provide a seal of quality for commercial composted products. Europe in particular enjoys a well-developed eco-labeling program for composted products.

The first criterion for the quality of compost is its stability or maturity. Although partially finished compost may have useful applications, fully mature compost is the most stable and uniform—and therefore valuable—form of the material. Nutrient content can vary considerably, depending on the nutrients of the feedstock, composting temperature, any leaching, and whether supplemental nutrients were added. It should only be necessary to conduct an analysis of nutrient content if the product is being sold, or you have some concern about how much of a crop's nutrient budget it can provide.

Texture should be fine and crumbly, as well as fairly uniform. If there are coarse undecomposed particles, it can be screened for a finer-textured product for applications such as potting mixes. Commercial compost standards include evaluation of the amount of nondegradable particles such as stones, bits of plastic or metal, as well as the presence of weed seeds.

Commercial compost makers also undertake laboratory testing for organic matter content, water-holding capacity, bulk density, particle size, and moisture content. Organic matter content should be in the range of 45–65 percent, the water-holding capacity greater than 100 percent, the bulk density 800–1100 pounds/cubic yard (dairy manure compost is typically 800 pounds/cubic yard), the particle size less than 1 inch, and the moisture content 30–50 percent. Tests of curing or stabilization can include an analysis of volatile organic acids (VOAs) present. The presence of compounds such as acetic, butyric, propionic, or lactic acid indicates partially anaerobic fermentation and therefore instability. In Europe, nitrogen immobilization or tie-up has been observed in some composts, mostly due to high C:N ratios. Salts tend to concentrate in finished composts, and for this reason the product should be tested for salinity before use in seedling mixes or potting soil.

Midwest Biosystems offers a grading system for compost quality ranging from A to D, with A being disease-suppressive. They will also test for sulfates and sulfides, pathogens, nitrogen forms, C:N ratio, pH, conductivity, redox potential, sodium, and moisture levels, as well as aerobic plate count and seed germination. Compost grades are assigned based on these tests. For compost to grade A it must contain between 600 to 900 ppm nitrates, no sulfides, have a pH of 7.0–8.1, and show a 70–100 percent seed-germination rate in pure compost.

Finally, Soil Foodweb, Inc. and other labs (see the listing in the "Resources" section) can evaluate the microbial population, both biomass and diversity of organisms, of a compost sample. This lab will assay its disease-suppressive qualities, based on direct observation, metabolic activity, and/or fatty acid content. Steve Diver provides a chart that lists minimum standards per gram of compost as follows:

---

### Minimum Levels of Microorganism Population per Gram

(Moisture: 50%–70%)
Active bacteria: 2–10 micrograms
Total bacteria: 150–300 micrograms
Active fungi: 2–10 micrograms
Total fungi: 150–300 micrograms
Flagellates: 10,000
Amoebas: 10,000
Ciliates: 50–100
Beneficial nematodes: 10–50

*Source:* Diver, "Biodynamic Farming and Compost Preparation."

---

Professional composters make use of all of these tests to varying degrees. Arnie Voehringer, who has made compost since 1976, notes that the concern for quality monitoring has changed markedly since he managed the farm at the Rodale Research Center. "Back then, if it smelled good it was good." Today, he devotes considerable effort to monitoring quality, and does germination tests as well as nutrient analyses. Mike Merner has his compost tested at the University of Massachusetts, and Mike Lombard is happy with the University of New Hampshire compost test, which includes analysis of nitrate vs. ammonium levels and C:N ratios. He also uses the Solvita® test. Karl Hammer is working with the UVM compost-testing program, which is still under development. He also has an in-house lab where every batch is tested for nitrate, potassium, sodium, conductivity, and pH. He has done some testing with Soil Foodweb, Inc., especially for his compost tea product.

## Evaluating Commercial Compost Products

There are numerous reasons for purchasing compost, even if you're making your own. If you buy compost, either in bulk or bags, examine it for the quality criteria discussed above, including the results of laboratory analyses performed regularly. First check into the closest operations, and talk to other growers who have used their product. Local municipalities, in particular, may be composting yard wastes and leaves, which they make available to the public at no or little charge. Biosolids compost from municipal treatment plants, while deemed suitable for agricultural use by EPA, are not permissible under the NOP.

You should definitely get a declaration of the feedstocks used in the product, as well as a description of the composting process. If it's for a certified organic operation and you don't wish to treat it according to the rules for raw manure use, you will also need documentation that it complies with the NOP requirements for compost (see chapter 9). Larger compost operations may participate in a quality-assessment program, such as the Quality Seal of Approval classification performed by Woods End Labs in collaboration with the Rodale Institute. This scheme recognizes six types of compost, with minimum standards or criteria for each: Seed Starter, Container Mix, Garden Compost, Topsoil Blend, Mulch, and Natural Fertilizer. Mike Lombard has participated in this program in the past, but finds it somewhat expensive for his scale of operation.

## Where, When, and How Much to Apply

High-quality finished compost is highly versatile—it can be applied to any crop at any time, in any amount. While it's a valuable source of nutrients, perhaps its primary value is as a soil inoculant, stimulating greater biological activity and release of nutrients for crops. Accordingly, the greatest benefit comes from spreading the available compost thinly over the most acreage, rather than concentrating it in a smaller area. Thicker applications of compost will improve soil structure and provide other benefits of increased humus content such as aeration, moisture retention, and buffering of soil pH.

Partially finished compost is often used for crops that require higher nitrogen, such as corn. Keep in mind that if you are certified organic partially finished compost must be treated like raw manure according to the regulations, unless it does not contain manure or other animal materials. Other crops, such as potatoes, do not tolerate undecomposed material well and may develop disease problems. Mike Lombard sells partially composted horse manure that is mainly bedding as a mulch for landscapers and flower gardeners.

Nutrient contributions of finished compost are usually estimated at the NPK equivalent of 1-1-1. This may, of course, be lower or higher, based on the nutrient content of the feedstock, loss to leaching and volatilization, and whether any mineral amendments were added. Although these percentages appear low, they reflect only the immediately soluble forms of nitrogen, phosphorus, and potassium. Once the compost is incorporated into the soil, with reasonable drainage and moisture conditions, additional nutrients are released through bacterial mineralization over the course of several growing seasons. Thus, if you seek to supply 100 pounds of immediately available N per acre from your compost application, you would apply about 5 tons per acre. Much higher application rates, in the range of 20–40 tons/acre, are commonly used for intensive vegetable crops, and to provide the optimum disease-suppression effect. Joey Klein, who farms on very sandy soil in central Vermont, reports having good results spreading 15 tons of compost per acre before transplanting leafy greens, noting that the results justify the cost. Very high applications (50 tons or more per acre) of manure-based compost can, however, contribute to nitrate pollution of groundwater.

Mike Merner uses his own compost on his small CSA operation, and spreads a one-inch layer over the whole vegetable acreage nearly every year. He estimates that this amounts to about 120 cubic yards, or about 60 tons per acre—and has not observed any fertility problems with this high a rate.

## Timing and Application Rates for Various Crops

### Vegetables and Small Fruits

Applying anywhere from 1 to 50 tons per acre is common for vegetable producers, depending on amount available, crop, and soil condition.

Incorporate compost in the spring, when forming vegetable beds or in planting furrows.

### Trees and Perennials

Apply compost in planting holes or furrows.

*Orchards.* Spread about an inch the whole diameter of the dripline in fall after the leaves have fallen—the amount needed per acre depends on spacing and size of the trees. If approximately one-third of each acre of fruit trees is covered with compost, it will amount to about 20 tons per acre. Mow or chop to incorporate and derive maximum disease-suppressing benefit. Young trees benefit from fertility boost in early midsummer.

*Perennials.* Top-dress in early spring or fall.

### Pasture and Forage

Apply in spring, just when the field is starting to green up (if the field is dry). Compost can also be applied following the first cut of hay; in the pasture following intensive grazing; and in the fall, following the last cut.

### Field Crops

Apply prior to planting, when incorporating green manure, or following harvest.

## Special Uses: Disease Control and Anti-Fungal Mixes

Compost generally serves to suppress plant pathogens because it contains a rich diversity of beneficial microorganisms that act against plant diseases in two ways: As a general suppressive agent, beneficial organisms fill the ecological niche on plant leaves or roots that would otherwise be occupied by pathogens. They also produce antagonists to specific disease organisms, such as antibiotics or predatory fungi. Some compost organisms produce enzymes that break down cellulose and chitin, which are primary constituents of many plant pathogens, and in the process kill the pathogens.

While the high temperature of thermophilic composting kills many disease pathogens, it also kills many beneficial organisms. It is, however, possible to inoculate compost with the beneficials. Some research indicates that regular reinoculation of compost with specific fungi and bacteria that are known biocontrol agents is necessary to achieve consistent levels of disease suppression. The primary key to encouraging biocontrol organisms is to use fully mature, well-stabilized compost. This means allowing finished compost to cure as long as possible before using it for potting mixes, and incorporating it into soil the season before planting crops. Sufficient moisture is also crucial—compost should have at least 40–50 percent moisture to support colonization by disease-suppressive bacteria and fungi.

A fair amount of research has been done to identify particular plant diseases that are controlled by compost, and to examine what recipes and methods are most effective for producing disease-suppressive composts. This information is much more fully developed for potting mixes and the nursery industry than it is for field crops. Dr. Harry Hoitink, a plant pathologist at Ohio State University, is among the foremost researchers in this area. Among the primary plant pathogens suppressed by compost organisms are *Pythium* and *Phytophthora spp.*, which cause damping off and root rots. European researchers have used compost extracts augmented with cultured beneficials to control mildew in sugar beet, *Botrytis cinerea* in strawberries, powdery mildew, and blight in potatoes. There has also been some work on culturing specific biocontrol agents to inoculate compost, especially to control *Rhizoctonia solani*. Mike Merner used to be in the landscaping business, and used his own compost as his primary disease-control strategy for various lawn problems.

The major microbial biocontrol players consist of bacterial taxa such as *Bacillus, Pseudomonas, Flavobacterium,* and *Pantoea spp.*, as well as actinomycetes, which include the antibiotic-producing *Streptomyces.* Predatory fungi play a major role in controlling pathogenic nematodes such as those responsible for root cyst disease. Fungal biocontrol agents such as *Trichoderma spp.* seem to thrive best in substrates high in lignocellulose, such as tree bark and sawdust. On the other hand, *Penicillium spp.* proliferates in substrates to which composted grape pomace has been added, a medium low in cellulose and high in sugar.

There are a number of commercial products containing beneficial, disease-suppressive organisms. These products include seed treatments, compost inoculants, soil inoculants, and soil drenches. Among the beneficial organisms available are *Trichoderma, Flavobacterium, Streptomycetes, Gliocladium spp., Bacillus spp., Pseudomonas spp.*, and others. While some significant increases in disease suppression have been found with these inoculants, it is wise to provide a good feedstock (i.e., fully mature compost), and observe the effectiveness of inoculated versus uninoculated compost in your own farm environment.

# eight

# Compost Tea and Other Brewed Microbial Cultures

Compost tea has become a hot topic in recent years. Commercial enterprises that sell both compost tea and compost tea makers have proliferated like flies in a manure pile. Compost teas, according to Steve Diver of ATTRA, provide soluble nutrients that promote a noticeable greening of crops. They also provide bioactive compounds that function as biostimulants and coat the plant surfaces with protective microorganisms.

Once upon a time, when referring to compost tea you meant the liquid obtained from soaking a burlap sack of compost in a barrel of water. No longer. This substance is more properly considered a compost watery extract, whose primary benefit is as a supply of soluble nutrients. While useful, compost (or manure) extract does not confer the benefits of a true brewed compost tea. Compost tea goes beyond an extract by combining the compost with a microbial food source—usually added carbohydrates and minerals—and then extracting and growing populations of beneficial microorganisms through an aerobic process. The resulting microbe culture is then sprayed on foliage or used as a soil drench.

Materials other than compost may also be used to brew fertility-enhancing teas. These preparations, becoming more widely known as *indigenous* or *effective microorganisms*, are used in much the same way as compost tea. In some cases they use substrates similar to compost feedstocks, as well as various herbs and other plant materials, and may include some compost as an inoculant.

## Why Compost Tea Is Valuable

A rich brew of diverse beneficial microorganisms, applied directly to leaf surfaces (phyllosphere) or the root zone (rhizosphere), provides a certain amount of directly available plant nutrients. This is a practice well known to organic growers who apply foliar fertilizers such as fish emulsion and seaweed. But foliar fertilization is a way to give plants a direct shot of nutrients and circumvent one of the fundamental precepts of organic farming: *Feed the soil, not the plant.* Compost tea goes beyond even feeding the soil by establishing and enhancing microbial populations that, in addition to cycling more soil nutrients, suppresses pathogens.

The guru of compost tea is undoubtedly Dr. Elaine Ingham, president of Soil Foodweb, Inc., in Oregon. Dr. Ingham, who speaks and writes prolifically about the soil food web, works with conventional farmers and damaged soils to culture and restore beneficial microbial populations. According to Dr. Ingham, we are aware of only a tiny fraction of the beneficial species of microbes in soils, and even those that have been studied are often missing from damaged soil. She laments, "Soil scientists often tell me that the bacteria and fungi just 'return to the soil—rapidly.' But they don't. The residues in soil, the continued plowing, the lack of organic matter in ag soil, all contribute to even the few beneficial species we know about not being in most ag soils. We need to put them back. Compost and compost tea are excellent ways to do that" (SANET posting 10/27/02).

The concept behind compost tea and other beneficial microbial cultures is to provide a high level of these organisms directly to plant leaves and roots, where they will colonize and not allow pathogens to gain a foothold. Ingham states that, when the tea has a sufficient concentration of diverse microbial species, "the disease organisms have no place to grow, no way to infect the plant, no foods to eat because the beneficials already ate them before the disease had a chance, and you build soil structure so anaerobic conditions do not occur in your soils."

## How to Make It

As mentioned, the old burlap bag filled with compost and suspended in a bucket, while useful as a source of quick foliar nutrients, is a watery compost extract, not the "real deal" of fermented compost tea. To make a true fermented compost tea, you need some form of equipment and means of aeration. Compost tea brewers are becoming widely available, at widely varying costs (see the "Resources" section), but a homemade system can be created with less expense.

The trick to brewing compost tea is to extract the beneficial compost organisms from the compost particles and then culture them in a water solution. This requires time (at least twenty-four hours), aeration, and the addition to the water of a nutrient source for the microbes. Dr. Ingham describes the process thus:

> Compost tea machines basically take a very small amount of your good, aerobic (non-stinky, properly heated or processed by worms) thermal compost or worm compost, use water movement through the compost (air bubbling through, or pump the tea through), to extract the organisms from the compost by ripping them off the surfaces of the compost to get them into the water. Bacteria are well-glued to compost surfaces, and fungi are wound around the compost particles, like the threads they are, so they have to have a fair amount of force applied to get them off the surfaces. That's why dripping water through compost doesn't do much—the organisms don't want to get washed off their food. It's like trying to get starving people away from the dining table—not happening unless force is used. . . . Once you start the extraction process, then you want the beneficial organisms to grow in the water. Soluble nutrients from the compost are extracted, and that helps grow beneficials. Addition of humic acids, fulvic acids, small amounts of molasses, complex sugars, proteins, help grow good guys. Fish hydrolyzate grows good guys too, but you need to be careful of those materials. . . . You need to aerate well enough to keep the brew aerobic. If the tea drops into the anaerobic zone, you lose your fungi, and lose the ability to prevent mildew, anthracnose, blight, wilt, curl and many insects, like aphids, white flies, beetles, etc., from growing. And, if you let the tea go anaerobic, you are likely going to grow human pathogens if your compost was poor and not properly made, and had *E. coli* in it. (SANET posting 1/5/03)

## Brewing Methods

Steve Diver describes some alternative methods for making your own compost tea brewer. The first is the bucket-bubbler method, which begins with the burlap bag in the bucket and adds an aquarium size pump and air bubbler, along with some kind of microbial food and mineral sources. You may need to use multiple pumps to ensure that aeration is thorough enough. After two or three days the product will be ready and should be used immediately. With several buckets, a fresh batch can be available on a regular basis.

The trough method can be used to make larger quantities with home-made equipment. A length of 8- or 12-inch-diameter PVC pipe is cut in half, drilled full of holes, and lined with burlap, then suspended over a tank such as a horse trough or stock waterer. Compost is placed in the PVC-pipe trough and the lower tank is filled with water and amended with microbial nutrients. A sump pump in the tank moves the solution through a trickle line running the length of the suspended trough. The solution trickles through the compost, creating a leachate that drips a few feet through the air back into the tank. This process takes about seven days of recirculation, bubbling, and aeration.

There are now a fair number of commercial tea brewers, along with prepackaged microbial nutrient sources and other goodies, offered by at least ten different companies. A description of some of these companies and the types of equipment and supplies they offer is provided by Steve Diver in the ATTRA publication, "Notes on Compost Teas." In the Northeast, North Country Organics (see the "Resources" section) offers a compost tea brewer along with a custom product made by Karl Hammer for compost tea. Dr. Ingham emphasizes that the most important thing to look for is research data showing that the brewer does an adequate job of maintaining aeration throughout its cycle, and that the brew is regularly monitored for numbers and diversity of microbial populations. Ease of cleaning is also important, especially if batches are made sporadically rather than continually. She states that, "The tea machine maker should also be able to show you repeated trials where they measured total and active bacterial biomass, and total and active fungal biomass numbers in their teas, and that these figures should be within the range characterized as 'desired' by Soil Foodweb, Inc., test results" (SANET posting 1/5/03).

Otherwise, no significant differences in product quality have been observed in trials conducted by Soil Foodweb, Inc., comparing different commercial compost tea brewers, as long as they were used according to manufacturers' directions.

## Using Compost Tea

Once the tea is ready it should be used as soon as possible. Use it full strength as a foliar spray, or add it to a tank for watering-in transplants. A filtering system is essential to avoid clogging sprayers. If your soil already has a high level of biological activity, one dose early in the season will be enough to stimulate disease-suppressive organisms. It is important to make sure that you get good leaf coverage when spraying, to prevent colonization by plant pathogens. Depending on the conditions, additional applications through-out the growing season can give a boost to certain vulnerable crops.

A commercial-style compost tea brewer with pipe for aeration and spigot at bottom to drain tea out.

## Evaluating Quality

Compost tea should contain both an abundance and diversity of beneficial microorganisms. Dr. Ingham has pioneered the use of the *direct look* method, as opposed to plate counts, for assessing these qualities, in which she views and counts microorganisms with high-performance light microscopy enhanced with staining and illumination. Steve Diver gives minimum standards of microorganisms for compost and compost tea in "Notes on Compost Teas," including figures for mass of active and total bacteria, active and total fungi, flagellates, amoebas, ciliates, and beneficial nematodes. Fungi are critical in achieving the desired coverage and microbial balance; for this reason, hot composting temperatures, which suppress fungi, result in poor-quality compost tea. Soil Foodweb, Inc., offers testing services to evaluate the population and diversity of beneficial organisms in a given batch of compost or compost tea.

## Compost Tea and the NOP

"It is an unfortunate turn of events," says Steve Diver, "that NOP went down the path of *E. coli* hysteria at the same time the compost tea industry is taking off like wildfire, providing a dynamic new tool for organic agriculture." Unless you can document that the compost used was produced according to the NOP requirements (see chapter 9), the NOP treats compost tea as if it were raw manure. This means having to wait 120 days after application before you can harvest crops whose edible portions are sprayed, effectively kicking compost tea out of the life cycle of the crop. "Yet it is the inoculation of the leaf surface and the root surface with compost tea which results in such a dynamic benefit to the crop, especially at crucial stages of growth" (Diver, SANET posting 10/31/02). Even under this constraint, compost tea could be used early in the season (90 days before harvest) on fruit trees and other crops whose edible portions aren't contacted by the brew, which is often a time when diseases like scab are getting established.

While some claim that *E. coli* will not proliferate in properly made compost tea, there is no research that demonstrates that it is totally eliminated in a mixed culture. One researcher (Joel Reiten, SANET posting

11/4/02) has suggested that the best course of action for the NOP would be to require that compost or compost teas used within 120 days of harvest must have a documented negative *E. coli* test. Simple and rapid tests for *E. coli* are available and could be included as part of the organic farm plan, with documentation that they were being conducted when using a batch of compost tea. Various private and state laboratories offer inexpensive *E. coli* testing, and test kits may be obtained at a cost of less than $3 per sample.

Because the NOP rules apply specifically to animal materials, one way around them would be to work with plant-based fermented teas that culture effective or indigenous microorganisms.

## Effective/Indigenous Microorganisms

While not strictly speaking a composting process, the production and uses of indigenous or effective microorganisms has many similarities to the process of making and using compost tea. Effective microorganisms, or EM, is a mixed culture of beneficial microorganisms (primarily photo-synthetic and lactic acid bacteria, yeast, actinomycetes, fermenting fungi) that can be applied as an inoculant to increase the microbial diversity of soils. The technology behind the concept of effective microorganisms and its practical application was developed by Professor Teruo Higa at the University of the Ryukyus in Okinawa, Japan. This basic method can be used to produce cultures specifically designed to enhance particular crops, as well as cultures beneficial for both livestock and human nutrition.

Gil Carandang of the Philippines, in his booklet about indigenous microorganisms, suggests that the best way to promote healthy, diverse microbial life in soil is to ferment a culture of organisms that are specific to the plants you are growing. Microbial cultures will also produce what he refers to as *bionutrients*. The basic approach is to ferment a chopped mass of plant material (about a pound) with ⅓ cup of raw sugar or molasses and add enough water to cover. Put the materials in a container with at least a 50–75 percent air gap and cover loosely. After a week, molds and some alcohol will form, and this should be strained and diluted by 20:1. The resultant bionutrient is added to water at a rate of 2–4 tablespoons per gallon. This extract can be added to animal feed and drinking water in addition to using it to inoculate your compost pile or as a foliar spray or root drench at transplanting.

Carandang also discusses recipes for what he calls *designer compost*, known in Japan as *bokashi*. It consists of materials in the proportions of 80 percent carbon source, 17 percent nitrogen, and 3 percent trace elements. To 100 kilos of this mixture is added about a liter of indigenous microorganism brew, a liter of bionutrients, and a kilo of molasses. Nutrients can be tailored to your needs by using bionutrients made from materials high in a particular nutrient, such as fish wastes for nitrogen or alfalfa meal for potassium. Rock powders can also be used in the mix. Other recipes he provides include a bioavailable calcium phosphate produced from eggshells, and a fermented banana-squash-papaya extract that is especially helpful for stimulating flowering and fruiting.

# nine

# Compost and the Law

## NOP Requirements: What Do They Say, and Why Do They Say It?

Farmers certified organic under the National Organic Program (NOP) must comply with its requirements for use of raw manure as well as the rules that specify how compost must be made. This means that compost made in windrows has to be turned at least five times, and remain at a temperature between 131°F and 170°F for fifteen days. Static aerated or in-vessel systems must reach those temperatures for at least three days. Besides actually making the compost this way, you have to keep records, such as field logs, to document that you have done so.

Any compost that does not meet these criteria has to be treated like raw manure under the regulations, which means that it cannot be applied to crops for human consumption within the same growing season (90 days previous for crops whose edible portions are not contacted by soil or soil particles, and 120 days previous for crops whose edible portions are contacted by soil or soil particles). This only applies to compost that contains any animal materials; plant material, composted or uncomposted, is permitted with no extra restrictions. Compost purchased from off-farm sources must also meet these requirements as long as it contains animal matter. This creates a real handicap for those who are satisfied with the "let it happen" composting method mentioned in chapter 1, or who are using vermicompost systems in which high temperatures are detrimental. It also creates problems for specialized disease-suppressing compost formulas, which are best made at low temperatures. Use of manure tea is definitely prohibited under the NOP, and use of compost teas is under review by NOSB.

The rules for management of raw animal manure cover compost tea, according to the National Organic Program. As such, it can be used at least 90 days before harvest on fruit trees and other crops whose edible portions aren't contacted by the brew—a time when diseases like scab are getting established.

## Why Were the Rules Made So Restrictive?

The NOP compost requirements were taken directly from the Natural Resources Conservation Service's (NRCS) Conservation Practice Standard for a Composting Facility (Code 317). This practice standard applies to on-farm composting facilities, and is intended to ensure that they do not contribute to water pollution or create other environmental hazards from human pathogens, dust, and odors. They were not intended as a guide to producing the best compost possible. At the time the NOP regulations were being developed there was substantial public concern about human pathogens, particularly some deadly *E. coli* O157:H7 and *Salmonella* outbreaks in meat, fresh berries, juices, and other foods. Some of this contamination came from livestock manure, and so the explicit allowance of raw animal-manure use in the enabling federal Organic Food Production Act and the regulations provoked alarm about possible food safety concerns in organic food. This was reflected in several public comments, some of which suggested that organic farmers should never be allowed to use raw animal manure. It also became clear that the definition of compost was not (and many would argue, could not be) precise enough to ensure that it would be made in a way that eliminated any human pathogens. In addition, the issue became politically charged by public attacks on organic farming by Dennis and Alex Avery of the right-wing Hudson Institute, which reached its peak in a segment of ABC's *20/20* in which it was inaccurately claimed that organic foods pose a higher health risk from pathogens because of organic farmers' use of manure.

Although this concern was clearly based on false information and reasoning (no such restrictions are placed on conventional farmers who use manure), the rule was made this restrictive to deflect any such alarmist claims and to gain the approval of the various government agencies that had to sign off on the regulation. There was no consideration in this scheme for questions about compost quality or the practicality of the rules for farmers.

## What Alternatives Are There?

If you are committed to marketing your products under an organic label, you have no alternative but to comply with the regulations as currently written. However, regulations can change—the law requires only that manure be used in such a way that food safety is not compromised. When this regulation was proposed, the preamble stated that "this requirement is especially challenging given that there is no Federal oversight of the application of raw manure to any kind of crop nor any public health standards to establish what constitutes safe use of raw manure" (FR V.65, No.49, 3/13/2000, pg. 13540). Subsequently, the NOSB (National Organic Standards Board) Crops Committee created a Compost Task Force and issued recommendations that elaborated on and adjusted this rule to provide greater flexibility. In addition to specifying the proper conditions for vermicomposting, it recommends that, in the case of wind-rowed compost, "(*ii*) the compost undergoes an increase in temperature to at least 131°F (55°C) and remains there for a minimum of 3 days, and (*iii*) the compost pile is mixed or managed to ensure that all of the feed-stock heats to the minimum temperature" (NOSB Compost Task Force Final Report). Although the vermicomposting recommendation was not adopted, the final rule does allow for static or in-vessel systems.

In October 2010 The National Organic Program published additional guidance documents intended to implement previous NOSB recommendations, including one entitled "Compost and Vermicompost in Organic Crop Production." Following an opportunity for public comment, this guidance is expected to be included in the Program Handbook, and specifically includes the NOSB Compost Task Force recommendations identified above as examples of acceptable composting and vermicomposting methods. The methods used for composting and vermicomposting will have to be described in the farmer's Organic System Plan, and can include performance measures, such as temperature, time, moisture content, chemical composition, biological activity, and particle size to document compliance (http://www.ams.usda.gov/AMSv1.0/getfile?dDocName=STELPRDC5086966&acct=nopgeninfo).

In the meantime, if you need organic certification, you will either have to obtain documentation that any compost you buy has been produced according to the NOP rules or be able to document your own compliance in producing compost on your farm. A sample log form is provided (see appendix E) to help you do this. Alternatively, you could treat any manure-based compost as if it were raw manure and use it for crops for human consumption only at least 120 days before harvest of crops whose edible portions may be contacted by soil particles (such as most vegetables), or 90 days before harvest of crops whose edible portions do not contact soil particles (such as tree fruits).

All the professional composters interviewed for this manual make their compost in compliance with the NOP, and can document this compliance. Most feel that the regulations make some sense, but are too rigid about things like requiring it to be turned at least five times. Karl Hammer addresses this by sometimes turning twice in the same day, rather than setting up his equipment that many times throughout the season.

## State Regulations

Compost operations larger than garden-scale may have to comply with state and local regulations concerning environmental quality, health, and odor control, especially in areas with large nonfarm populations. These requirements are generally measures to protect groundwater from nutrient runoff and contaminants such as pathogens, to protect the public from health threats due to vermin, flies, and other harmful organisms, and to guard against obnoxious smells downwind. Your local cooperative extension or state environmental agency should be able to provide you with information about compliance with compost facility regulations. Highfields Center for Composting can also assist with site permitting and regulatory review.

There are several compost industry organizations (see the "Resources" section) that may also be able to help you better understand the regulatory environment, provide technical assistance in overcoming regulatory barriers, and lobby for the interests of environmentally friendly waste-recycling operations. According to one western waste-recycling consultant,

> Many states are beginning to look at performance-based rather than prescriptive regulations. It is an unfortunate irony that in some states, composting [tree] leaves requires lengthy environmental review while hundreds of tons of manure pile up unmanaged and unregulated. Many folks believe that with the reauthorization of the Clean Water Act, the EPA Confined Animal Feeding Operation regulations, and other proposed regulations, the increased regulation of manure is just down the road. The incentives for less regulation would be a demonstration of reasonable performance from the composting industry. . . . Several progressive states are taking a hard look at existing regulations and trying to ease requirements for well-operated facilities while maintaining some oversight over the bad operators. (Matt Cotton, "Compost Education & Resources for Western Agriculture," Integrated Waste Management Consulting, San Francisco.)

There are also regulatory issues to consider when it comes to marketing finished compost. Jurisdiction over how fertilizer products must be labeled, their guaranteed analysis, and similar issues reside with each individual state. The U.S. Composting Council has been working with the American Association of Plant Food Control Officials (AAPFCO) to come up with suggested rules for compost quality. Concerns about this issue stem from information about high content of heavy metals and other contaminants in some industrial waste products used as fertilizers. The EPA currently has rules for categorizing biosolids (otherwise known as sewage sludge) according to the product's suitability for various applications, based on its content of heavy metals and other contaminants. At this point the decision about formal regulation of commercial compost products remains up to each individual state.

Mike Merner reports that among his biggest problems as a marketer of compost is the competition from municipal composting operations that give away their product at no charge, though they offer poorer quality (i.e., they contain litter and debris and are unscreened). He complies with all local regulations, which have escalated significantly in recent years, imposing a serious burden on a small-scale operation. He says, "It's much harder to get started in this business today because of the regulatory environment."

On-farm composting may be subject to different, and generally less stringent, regulations than stand-alone facilities. In most states requirements for composting your own manure fall under general farm regulations and do not require additional permits. While many states are moving to performance-based regulations, many are simultaneously using feedstocks, and their associated risks, as a basic regulatory framework. Since food scraps are commonly regulated as a solid waste, you will often require state solid-waste permits to compost food scraps. Karl Hammer is able to get around this by feeding the food scraps to his chickens first, so it is technically still just a farm.

On the helpful side, USDA's Natural Resource Conservation Service (NRCS) also provides some cost-share for manure management systems, including composting. Many state agencies also supplement this cost-share, which is generally only for livestock operations.

## Conclusion: The Future of Composting

Although the increased regulation of composting operations presents new challenges, it is also an indication of how greatly composting has proliferated over the past decade. As is true of the regulation of organic production in general, widespread adoption of composting by municipalities, farmers, and food processors means that more organic wastes are being recycled to return their nutrients to the soil, and more land is being enriched with organic matter instead of synthetic fertilizers, with reduced environmental hazards from landfills and nutrient runoff. A technology that was once considered unscientific or silly is now an important and expanding option with proven benefits to soil and plant health as well as the bottom line.

# Appendix

## A. Recipe Calculators

### Co-Composter

Cornell's Department of Biological and Environmental Engineering and Waste Management Institute have developed Co-Composter, an Excel spreadsheet model for the planning of co-composting systems for mixtures of dairy manure and other organic wastes. Co-Composter provides mass and volume balances, area estimations, and a cost analysis of alternate composting systems based on inputs entered by the user. Co-Composter is available at www.compost.css.cornell.edu/CoComposter.html.

### Calculate C:N Ratio for Three Materials

To calculate the carbon-to-nitrogen ratio for up to three materials, you can enter the wet weight, percentage of carbon, percentage of nitrogen, and percentage of moisture on a Web page and click on the calculate button at: www.compost.css.cornell.edu/calc/2.html.

### Compost Mixture Calculation Spreadsheets

You can download spreadsheets with built-in equations to solve compost-mixture calculations for up to four ingredients, from: www.compost.css.cornell.edu/download.html.

## Recipe Formulas

| Individual Ingredient Formulas: | |
|---|---|
| Moisture content | % moisture content/100 |
| Weight of water | Total weight × moisture content |
| Dry weight | Total weight − weight of water<br>Total weight × (1 − moisture content) |
| Nitrogen content | Dry weight × (% N / 100) |
| Percent carbon | % N × C:N ratio |
| Carbon content<br>(N content × C:N ratio) | Dry weight × (% C / 100) |

**General Mix Formulas:**

$$\text{Moisture content} = \frac{\text{Weight of water in ingredient a + water in b + water in c + ...}}{\text{Total weight of all ingredients}}$$

$$\text{C:N ratio} = \frac{\text{Weight of C in ingredient a + weight of C in b + weight of C in c + ...}}{\text{Weight of N in a + weight of N in b + weight of N in c + ...}}$$

# B. Sample Compost-Production Budget Sheet

| 1. Tasks: | | | |
|---|---|---|---|
| Task | Monthly labor (hours) | Monthly machine time (hours) | Comments |
| Site preparation (initial expense) | | | |
| Loading and trucking materials | | | |
| Pile formation, incl. chopping and mixing materials | | | |
| Pile turning | | | |
| Screening | | | |
| Field spreading | | | |

## 2. Materials:

| Compostable materials needed / avail. | Quantity | Labor needs income | Cost or timing | Notes (sources, special handling, etc.) |
|---|---|---|---|---|
| **On-farm** | | | | |
| Manure | | | | |
| Spoiled hay | | | | |
| Vegetable trimmings | | | | |
| **Off-farm** | | | | |
| Sawdust | | | | |
| Cardboard | | | | |
| Cannery wastes | | | | |

## 3. Equipment:

| Equipment | Model/ descrip. | Purchase cost | Operation cost | Notes (age, repair needs, etc.) |
|---|---|---|---|---|
| Tractor | | | | |
| Loader | | | | |
| Shredder | | | | |
| Manure spreader | | | | |
| Truck | | | | |
| Hand tools | | | | |
| Monitoring equipment | | | | |
| Aeration system | | | | |
| Supplies (worms, additives) | | | | |

# C. Potting-Mix Recipes

The following recipes were taken from the ATTRA publication, "Organic Potting Mixes for Certified Production," by George Kuepper and Katherine Adam, February 2002.

### Classic soil-based mix

⅓ mature compost or leaf mold, screened

⅓ garden topsoil

⅓ sharp sand

*Note:* This mix results in a potting soil that is heavier than modern peat mixes, but still has good drainage. Vermiculite or perlite can be used instead of sand. Organic fertilizer can be added to this base.

### Organic substitute for Cornell Mix

½ cu. yd. sphagnum peat

½ cu. yd. vermiculite

10 lbs. bonemeal

5 lbs. ground limestone

5 lbs. bloodmeal

### Seedling mix for Styrofoam seedling flats

2 parts compost

2 parts peat moss

1 part vermiculite, pre-wet

### Sowing mix

5 parts compost

4 parts soil

1–2 parts sand

1–2 parts leaf mold, if available

1 part peat moss, pre-wet and sifted

*Note:* All ingredients sifted through a ½-inch screen. Add 2 Tbsp. lime for every shovelful of peat to offset the acidity.

## Prick-out mix for growing seedlings to transplant size

6 parts compost
3 parts soil
1–2 parts sand
1–2 parts aged manure
1 part peat moss, pre-wet and sifted
1–2 parts leaf mold, if available
1 six-inch pot bone meal

## Special potting mix

1 wheelbarow load sifted soil
1 wheelbarrow load aged manure
1 wheelbarrow load sifted old flat mix
5 shovelfuls sifted peat
2 four-inch pots bone meal
2 four-inch pots trace mineral powder
2 four-inch pots blood meal

## Sterile peat-lite mix

½ cu. yd. shredded sphagnum peat moss
½ cu. yd. horticultural vermiculite
5 lbs. dried blood (12%N)
10 lbs. steamed bone meal
5 lbs. ground limestone

## Recipe for soil blocks

20 qts. black peat
20 qts. sand or calcined clay
20 qts. regular peat
10 qts. soil
10 qts. compost
½ cup lime
1 cup colloidal phosphate
1 cup blood meal
*Note:* All bulk ingredients should be sifted through a ½-inch screen.

### Organic potting mix

    1 part sphagnum peat

    1 part peat humus (short fiber)

    1 part compost

    1 part sharp sand (builder's)

To every 80 qts. of this add:

    1 cup greensand

    1 cup colloidal phosphate

    1½ to 2 cups crab meal or blood meal

    ½ cup lime

### Blocking-mix recipe

    3 buckets (10-qt.) brown peat

    ½ cup lime

    2 buckets coarse sand or perlite

    3 cups base fertilizer (blood meal, colloidal phosphate, and greensand mixed together in equal parts)

    1 bucket soil

    2 buckets compost

### Blocking mix for larger quantities

    30 units brown peat

    ⅛ unit lime

    20 units coarse sand or perlite

    ½ unit base fertilizer (blood meal, colloidal phosphate, and greensand mixed together in equal parts)

    10 units soil

    20 units compost

### Mini-block recipe

    16 parts brown peat

    ½ part colloidal phosphate

    ½ part greensand (leave it out if not available—do not substitute a dried seaweed product)

    4 parts compost (well decomposed)

## ATTRA Recipe #6

40 qts. sphagnum peat moss

20 qts. sharp sand

10 qts. topsoil

10 qts. mature compost

4 oz. ground limestone

8 oz. blood meal (10% N)

8 oz. rock phosphate (3% P)

8 oz. wood ashes (10% K)

## Disease-suppressive mix

50% or more light (as opposed to highly decomposed dark) fibrous sphagnum peat

20% or more composted pine bark

enough lime to produce a pH of 5.3–5.5

*Note:* Age this mix, with 40%–45% water content, for a couple of weeks at temperatures between 65°F and 90°F before planting, to allow beneficial organisms to become well established.

# D. Current Compost Regulations of the NOP

## Subpart A: Definitions:

**Compost:** The product of a managed process through which microorganisms break down plant and animal materials into more available forms suitable for application to the soil. Compost must be produced through a process that combines plant and animal materials with an initial C:N ratio of between 25:1 and 40:1. Producers using an in-vessel or static aerated pile system must maintain the composting materials at a temperature between 131°F and 170°F for 3 days. Producers using a windrow system must maintain the composting materials at a temperature between 131°F and 170°F for 15 days, during which time, the materials must be turned a minimum of five times.

### Section 205.203(c):

The producer must manage plant and animal materials to maintain or improve soil organic matter content in a manner that does not contribute to contamination of crops, soil, or water by plant nutrients, pathogenic organisms, heavy metals, or residues of prohibited substances. Animal and plant materials include:

(1) Raw animal manure, *which must be composted* unless it is:

(*i*) Applied to land used for a crop not intended for human consumption;

(*ii*) Incorporated into the soil not less than 120 days prior to the harvest of a product whose edible portion has direct contact with the soil surface or soil particles; or

(*iii*) Incorporated into the soil not less than 90 days prior to the harvest of a product whose edible portion does not have direct contact with the soil surface or soil particles;

(2) Composted plant and animal materials produced through a process that:

[definition of compost]

(3) Uncomposted plant materials.

## E. Sample Logs to Document Compliance

These forms may be used to document compliance with NOP compost requirements in Section 205.203(c).

### Sample Compost Supplier Affidavit

Composted product brand name: _____

Supplier: _____

(company that made the compost)

     I, _____ (name of authorized agent of supplier) hereby certify that this composted product was produced according to the requirements of 7 CFR Part 205, Section 205.203(c)(2), so that:

(*a*) raw materials used to produce the composted product consist only of organic plant or animal materials, none of which have been chemically altered by a manufacturing process, along with mineral or bacterial additives that are permitted according to 7 CFR Part 205, Sections 205.601–205.602,

(*b*) the materials were combined to establish an initial Carbon to Nitrogen ratio of between 25:1 and 40:1,

(*c*) the materials were composted using a static or in-vessel system that maintained a temperature of between 131°F and 170°F for 3 days; or

(*d*) the materials were composted using a windrow system, which was turned at least 5 times and maintained a temperature of between 131°F and 170°F for 15 days.

Agent_____

Date_____

Company name & address _____

_____

| Compost Windrow Management Log | | | |
| --- | --- | --- | --- |
| Windrow ID or location: _____ | | | |
| Date | Activity | Temperature | Notes |
| | Pile built<br>Pile turned<br>Pile moistened<br>Monitoring activities<br>Removed for use | | (Materials used, C:N ratio, size, weather, precipitation, odors, color, other observations) |

# Resources

## Books and Publications

The following print and Internet publications are useful references for additional details or current information about composting.

*BioCycle*, Emmaus, PA: JG Press, Inc. biocycle@jgpress.com, www.jgpress
.com.
*BioCycle* is the leading trade magazine on farm, municipal, and industrial composting, published since 1960. JG Press also publishes numerous books and reports on composting, including *BioCycle Guide to the Art and Science of Composting*, 1991.

*Compost Science and Utilization*, Emmaus, PA: JG Press, Inc. http://www
.jgpress.com/compostscience.
A quarterly peer-reviewed journal focusing on management techniques to improve compost process control and product quality, with special emphasis on utilization of composted materials.

Chiumenti, Alessandro, and Roberto Chiumenti, Luis F. Diaz, George M. Savage, Linda L. Eggerth, and Nora Goldstein. *Modern Composting Technologies.* http://www.modern-composting.com.

Dougherty, Mark, ed. 1999. *Field Guide to On-Farm Composting.* NRAES 114. Ithaca, NY: Northeast Regional Agricultural Engineering Service.
Developed to assist in day-to-day compost-system management, spiral bound and printed on heavy paper with a laminated cover.

Haug, R. T. 1993. *The Practical Handbook of Compost Engineering.* Boca Raton, FL: Lewis Publishers.
A 717-page technical reference with lots of engineering and detailed specifications, which should be available at select libraries.

Koepf, H. H., B. D. Petterson, and W. Schaumann. 1976. *Bio-Dynamic Agriculture: An Introduction.* Spring Valley, NY: Anthroposophic Press.
The basic text on biodynamic farm practices and theory.

Lampkin, Nicholas. 1990. *Organic Farming*. Ipswich, UK: Farming Press
   Books.
A comprehensive and authoritative work on all aspects of organic
agriculture.

Martin, Deborah, and Grace Gershuny, eds. 1992. *The Rodale Book of
   Composting*. Emmaus, PA: Rodale Press.
The most recent update to the Rodale classic, primarily oriented toward
home gardeners.

NRCS Soil Quality Institute.*The Soil Biology Primer*. www.soils.usda.gov/
   sqi/soil_quality/soil_biology/soil_biology_primer.html.
Downloadable text, or printed version can be ordered.

Rynk, Robert, ed. 1992. *On-Farm Composting Handbook*. NRAES 54.
   Ithaca, NY: Northeast Regional Agricultural Engineering Service.
At 186 pages, the most comprehensive publication available on the subject,
with lots of technical details and specifications.

*Worm Digest*. Eugene, OR: Edible City Resource Center. mail@worm
   digest.org, www.wormdigest.org.
A quarterly newsletter that covers the use of worms in composting.

## Organizations and Internet Sites

### Government and Extension Information and Publications
**ATTRA (Appropriate Technology Transfer for Rural Areas)**
(800) 346-9140
www.attra.org
An incredibly useful source of information about anything related to
sustainable and organic agriculture, and an invaluable reference point
for this publication. Among the publications that can be downloaded are
"Farm-Scale Composting Resource List," (updated in 2005 at http://
attra.ncat.org/attra-pub/farmcompost.html), "Compost Teas for Plant
Disease Control," "Notes on Compost Teas," "Worms for Composting
(Vermicomposting)," "Disease Suppressive Potting Mixes," "Organic
Potting Mixes," "Alternative Nematode Control," "Sources for Organic
Fertilizers and Amendments," and "Sustainable Management of Soil-Borne
Plant Diseases."

### Cornell University Composting Home Page

http://compost.css.cornell.edu/Composting_homepage.html
Downloadable information includes select pieces of the NRAES handbook,
"Agricultural Composting: A Feasibility Study for New York Farms";
"Cornell Composting Resources and Publications"; and "The Science and
Engineering of Composting." Also available is Co-Composter, an Excel
spreadsheet model for the planning of co-composting systems for mixtures
of dairy manure and other organic wastes.

### EPA Solid-Waste Management: Composting Resources

www.epa.gov.
Among the publications available from EPA is a series entitled, "Innovative
Uses of Compost," with items on "Disease Control for Plants and
Animals," "Erosion Control," "Turf Remediation and Landscaping," and
"Reforestation, Wetlands Restoration, and Habitat Revitalization."

### Northeast Regional Agricultural Engineering Service (NRAES)

www.nraes.org/
An interdisciplinary, issue-oriented program sponsored by cooperative
extensions of fourteen member land grant universities. There are currently
eight publications on composting, including "Composting to Reduce the
Waste Stream: A Guide to Small Scale Food and Yard Waste Composting"
(NRAES 43).

### USDA National Organic Program

www.ams.usda.gov/
NOP Regulations and NOSB Compost Task Force recommendations.

### State Agencies
### CalRecycle (California) page on Organic Materials

http://www.calrecycle.ca.gov/organics/
A series of downloadable publications on compost and yard waste is available, including case studies of various demonstration sites.

### Maine Compost School

http://www.composting.org
The objective of the Maine Compost School is to provide training to
people interested and/or involved with medium- and large-scale composting operations. This course is offered as a certificate program  and will train
personnel to be qualified compost-site operators.

**Summary of Northeast States' Composting Regulations (2010 update)**
http://www.nerc.org/documents/compost_fertilizer/summary_compost_and_fertilizer_regulations.pdf

## Educational and Trade Organizations
**Biodynamic Farming and Gardening Association**
25844 Butler Road
Junction City, OR 97448
(888) 516-7797
(541) 998-0105
(541) 998-0106 fax
info@biodynamics.com
www.biodynamics.com

**Composting Association of Vermont**
(802) 223-1903
info@compostingvermont.org
www.compostingvermont.org
A nonprofit organization to educate people about using and producing compost, and serve as an information resource for Vermonters.

**Composting Council of Canada**
www.compost.org
A national nonprofit, member-driven organization with a charter to advocate and advance composting and compost usage. It serves as the central resource and network for the composting industry in Canada.

**Highfields Center for Composting**
Thomas Gilbert, Executive Director
P.O. Box 503
Hardwick, Vermont 05843
(802) 472-5138
www.highfieldscomposting.org
Technical services for on-farm composting and comprehensive food-waste recycling programs, action-based environmental education, workshops, publications.

## HowToCompost.Org

http://www.howtocompost.org

Links to information on home composting, large-scale composting, vermicomposting, composting toilets, and products and services. Also the Composter's Forum bulletin board.

## Josephine Porter Institute for Applied Biodynamics

PO Box 133

Woolwine, VA 24185

(276) 930-2463

(276) 930-2475 fax

info@jpibiodynamics.org

www.jpbiodynamics.org

Source for Biodynamic preparations produced according to Steiner's instructions.

## Northeast Organic Farming Association

411 Sheldon Rd.

Barre, MA 01055

www.nofa.org

Workshops, videotaped presentations, and printed materials about organic agriculture.

## Northeast Recycling Council

139 Main Street, Suite 401

Brattleboro, VT 05301

(802) 254-3636

(802) 254-5870 fax

info@nerc.org

www.nerc.org

## Sustainable Agriculture Research and Education Program (SARE)

www.sare.org.

SARE's Web site is maintained by the USDA-funded Sustainable Agriculture Network (SAN) for the Sustainable Agriculture Research and Education (SARE) program, which works with producers, researchers, and educators to promote farming systems that are profitable, environmentally sound, and good for communities. Topical indexes allow the user to browse a variety of subjects, including animal production, crop production, and economics and marketing. Provides information on networking opportunities, including the SANET online discussion group, and funding sources for research and education initiatives.

## U.S. Composting Council
1 Comac Loop 14 B1
Rokonkoma, NY 11779
(631) 737-4931
(631) 737-4939 fax
http://CompostingCouncil.org/
The USCC is a national, nonprofit trade and professional organization promoting the recycling of organic materials through composting.

## Businesses and Consultants
### Champlain Valley Compost Company
(802) 425-5556
(802) 425-5557 fax
info@cvcompost.com
www.cvcompost.com
Dealer of Compostex covers.

### Highfields Center for Composting
See listing in previous section.

### North Country Organics
(802) 222-4277
(802) 222-9661 fax
ncinfo@norganics.com
www.norganics.com
Compost tea brewers and accessories, as well as a full range of approved organic fertilizers and soil amendments.

### Soil Foodweb, Inc.
Dr. Elaine Ingham
Corvallis, OR
(541) 752-5066
(541) 752-5142 fax
sfi@soilfoodweb.com
www.soilfoodweb.com
Information about compost and compost tea biology and testing.

**Robert L. Spencer**
Environmental Planning Consultant
15 Christine Court
Vernon, VT 05354
(978) 479-1450
spencebbc@aol.com

**Vermont Compost Company**
(802) 223-6049
www.vermontcompost.com

**Waste Not Resource Solutions**
Brian Jerose
1662 Pumpkin Village Road
Enosburg Falls, VT 05450
(802) 933-8789 or 933-8336
jerose@together.net
www.farmcomposting.com

**Woods End Agricultural Institute**
Will Brinton
Mt. Vernon, ME
(207) 293-2457
(207) 293-2488 fax
info@woodsend.org
www.woodsend.org
Complete compost testing service, as well as various publications about compost quality, testing, and costs, and Solvita® compost test kits.

## Compost Testing Services
**A&L Eastern Laboratories, Inc.**
7621 Whitepine Road
Richmond, VA 23237
(804) 743-9401
(804) 271-6446 fax
office@al-labs-eastern.com
http://al-labs-eastern.com

**BBC Laboratories, Inc.**
1217 North Stadem Drive
Tempe, AZ
(480) 967-5931
(480) 967-5036 fax
Bbc@aol.com
www.bbclabs.com

**Midwest Bio-Systems**
28933 35 E. Street
Tampico, IL 61283
(815) 438-7200
(800) 355-6501
(815) 438-7028 fax
mbs@midwestbiosystems.com
www.midwestbiosystems.com

**Soil Foodweb, Inc.**
See listing in previous section.

**University of Massachusetts Soil Testing Lab**
West Experiment Station
University of Massachusetts
Amherst, MA 01003-8021
(413) 545-2311
soiltest@umext.umass.edu
www.umass.edu/plsoils/soiltest/
Land grant colleges in Maine, Pennsylvania, and New Hampshire also offer
compost-testing services.

**Woods End Agricultural Institute**
See listing in previous section.

# References

Carandang, Gil A. "Indigenous Microorganisms—Grow Your Own: Beneficial Indigenous Microorganisms and Bionutrients in Natural Farming." EM Application Manual for APNAN Countries, Asia-Pacific Natural Agriculture Network. Calamba City, Laguna, Philippines: Herbana Farms, 2003. gil_carandang@hotmail.com.

Cotton, Matt. 2001. *Compost Education & Resources for Western Agriculture.* San Francisco, CA: Integrated Waste Management Consulting.

Dickerson, George W. *Vermicomposting Guide.* Publication H-164. College of Agriculture and Home.

Economics, New Mexico State University. www.cahe.nmsu.edu/pubs/_h/h-164.html.

Diver, Steve. 1999. "Biodynamic Farming and Compost Preparation." Publication #IP137. Butte, MT: ATTRA. http://attra.ncat.org/attra-pub/biodynamic.html.

Gershuny, Grace, and Joseph Smillie. 1999. *The Soul of Soil: A Guide to Ecological Soil Management.* 4th ed. White River Junction, VT: Chelsea Green Publishing Co.

Kuepper, George, and Katherine Adam. 2002. "Organic Potting Mixes for Certified Production," Butte, MT: ATTRA.

Lampkin, Nicholas. 1990. *Organic Farming.* Ipswich, UK: Farming Press Books.

Rosenow, Paul, and Michael J. Tiry. *Composting Dairy Manure for the Commercial Markets.* Extended abstracts of papers and posters presented at the Manure Management Conference February 10–12, 1998, in Ames, Iowa, sponsored by West Central Region Soil and Water Conservation Society.

SANET (Sustainable Agriculture Network), an electronic listserve devoted to the topic of sustainable agriculture. Postings by Elaine Ingham, Steve Diver, Gil Carandang, and others about composting, compost teas, effective microorganisms and compost biology have been consulted extensively.

Singh, Av. *Demystifying Compost: A Closer Look into the Pile.* Organic Agriculture Centre of Canada.

# Glossary

**Aerobic:** A process that requires the presence of free oxygen, or a condition in which free oxygen is present.

**Anaerobic:** A process or condition in which free oxygen is absent, or very limited.

**Biodynamics (BD):** A school of agricultural thought, based on the spiritual teachings of Rudolph Steiner, that views the living interrelationships of soil, farm, and community as an organic whole.

**Biosolids:** Sewage sludge that has been designated by EPA as safe for use on food crops. Not permitted under the National Organic Program (NOP).

**Carbonaceous:** Plant matter that has a high percentage of carbon in relation to nitrogen, and is generally dry and bulky.

**Chelating:** The process of forming compounds consisting of a metallic element bound within a complex organic molecule, such as chlorophyll.

**C:N ratio:** The proportion of carbon to nitrogen by weight in any organic matter. The optimum level for biological activity in raw organic matter is between 20:1 and 30:1.

**Feedstock:** Raw organic materials used to feed the microorganisms that produce compost.

**Fermentation:** Chemical change brought about by action of yeasts and fungi that transforms sugars into ethyl alcohol, under anaerobic conditions.

**GMOs (genetically modified organisms):** Any organism in which the genetic material has been altered in a way that does not occur naturally by mating and/or natural recombination. (An organism here means any microbiological entity, cellular or noncellular, capable of replication or of transferring genetic material.)

**Humus:** The fragrant, spongy, nutrient-rich material resulting from decomposition of organic matter.

**Humification:** The process whereby raw organic matter is transformed into humus.

**In-vessel system:** An industrial-scale composting or anaerobic fermentation system in which the process takes place entirely within an enclosed container.

**Inoculants:** Cultures of beneficial microorganisms that may be added to compost, used to coat seeds of legumes, or added in processes for food products such as yogurt or wine to produce the desired product quality.

**Leachate:** Water from rain or snowfall that has drained through compost or manure, containing soluble nutrients.

**Macrofauna:** Animal organisms that are large enough to see, or nearly so, such as insects, nematodes, mammals, etc.

**Mesophilic:** Adapted to moderate temperatures (50°F–115°F), as microorganisms and macrofauna in the early and late phases of hot composting.

**Mineralization:** The release of soluble minerals and simple organic compounds through the decomposition of organic matter.

**Nitrogenous:** Organic material that may be of plant or animal origin, having a low proportion of carbon to nitrogen, and which tends to be heavy, wet, and dense.

**NOP:** The National Organic Program, which is part of USDA's Agricultural Marketing Service, having the responsibility for regulating organic food production and labeling.

**Pathogen:** An organism capable of causing disease in another organism.

**Putrefaction:** The process of anaerobic decomposition.

**Thermophilic:** Adapted to higher temperatures (113°F–160°F), generally bacteria, found as organic matter heats up by the action of mesophilic organisms.

**Tipping fees:** Fees charged by solid-waste-disposal operations to pay for the cost of tipping dumpsters of waste materials.

**Vermicompost:** Compost created by the action of earthworms, composed largely of their nutrient-rich castings.

**Volatilization:** The escape of chemical elements, such as nitrogen and carbon, into the atmosphere after being transformed into a gaseous state.

**Windrow:** A row of material formed in a long, horizontal pile.

# Index

# About the Author and Illustrator

Grace Gershuny is nationally known in the alternative agriculture movement, having worked for over thirty years as an organizer, educator, author, and consultant as well as a small-scale market gardener. She has written several books and numerous articles on soil management and composting, including *The Soul of Soil*, coauthored with Joe Smillie and *Start with the Soil*, published by Rodale Press, and edited the most recent edition of *The Rodale Book of Composting*. She served as editor of *Organic Farmer: The Digest of Sustainable Agriculture* for its four-year existence.

Grace worked for NOFA in the 1970s and 1980s in many capacities, including developing its first organic certification program in 1977, and was a founding member of the Organic Trade Association in 1985. From 1994 to 1999 she served on the staff of USDA's National Organic Program, and was a principal author of its first, much-maligned proposed rule. She has taught organic gardening and food system issues for the Institute for Social Ecology for many years, and continues to serve as a graduate advisor for the Prescott College masters program in social ecology. Grace currently serves on the board of the Highfields Center for Composting and the ANSI Sustainable Agriculture Standards Committee. She consults for the organic industry, does some inspection work, and is writing a book about the real meaning of "organic." She still grows much of her own food at her homestead in Barnet, Vermont, where she lives with her partner, Peter Blose.

Jocelyn Langer is an artist, music teacher, and organic gardener, and the illustrator of the NOFA organic farming handbooks series. She illustrates and does graphic design work for alternative media and political events as well as organic-farming-related publications. Jocelyn lives in central Massachusetts.

Joey Klein was the special farmer-reviewer for this manual, and Abigail Maynard was the scientific reviewer.

"This logo identifies paper that meets the standards of the Forest Stewardship Council. FSC is widely regarded as the best practice in forest management, ensuring the highest protections for forests and indigenous peoples."

Chelsea Green is committed to preserving ancient forests and natural resources. We elected to print this title on 30% post-consumer recycled paper, processed chlorine-free. As a result, we have saved:

4 Trees (40' tall and 6-8" diameter)
1 Million BTUs of Total Energy
379 Pounds of Greenhouse Gases
1,823 Gallons of Wastewater
111 Pounds of Solid Waste

Chelsea Green made this paper choice because our printer, Thomson-Shore, Inc., is a member of Green Press Initiative, a nonprofit program dedicated to supporting authors, publishers, and suppliers in their efforts to reduce their use of fiber obtained from endangered forests.

For more information, visit www.greenpressinitiative.org

Environmental impact estimates were made using the Environmental Defense Paper Calculator. For more information visit: www.edf.org/papercalculator